10^{10} 10^9 10^8 10^7 10^6 10^5 10^4
1000
100
10
1 light year
10^5
10^4
10^3
100
10
1 AU
0.1
10^6
10^5
10^4 miles

The visible universe, from the Earth's core to the horizon, 14 billion light-years away.

First published 2005 by Walker & Co., New York

Published by Wooden Books Ltd.
8A Market Place, Glastonbury, Somerset

British Library Cataloguing in Publication Data
Tweed, M.
The Compact Cosmos

A CIP catalogue record for this massive book
is available from the British Library

ISBN 1 904263 42 9

100% recycled papers throughout.

Printed and bound in Shanghai, China
by Shanghai iPrinting Co., Ltd.
100% recycled papers.

THE COMPACT COSMOS

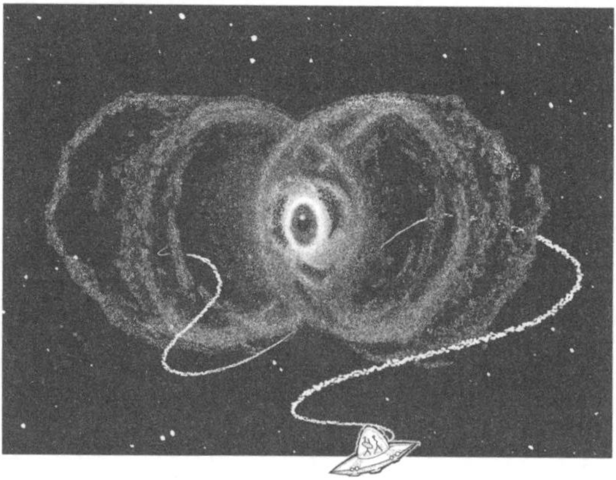

written and illustrated by

Matt Tweed

Thanks to H and the X-men

With love to Clair d'Lune and the Red Spider

For my family and friends

"Given enough blind men, elephant exploration would soon yield up secret after secret, and a new agreement could be forged in the community of blind men as to what an elephant is."

from *Ineffable Elephant* by Bert Speelpenning

CONTENTS

INTRODUCTION

As we gaze out into the vastness of space, or look within to the sub-atomic worlds of matter, an intricate tapestry is revealed. Vast harmonic weaves of energy criss-cross dimensions and, in passing, generate the space-time wonderland of the universe.

The essential paradox of existence has inspired generations to look to the skies and beyond. Even now, we are dipping our toes into a largely uncharted ocean that stretches far beyond our familiar, quiet corner of the Milky Way galaxy.

The science of cosmology is still at an early stage. Most of the universe is so far away, and behaves so bizarrely, that we have only the sketchiest understanding of the processes involved. This little book explores some of the ideas painstakingly developed by many great minds attempting to understand the magnificence.

At a point in time where humanity is waking up to the enormous changes happening to our own tiny planet, the heavens are revealing spectacular scenes of star-birth and calamity, splendor on a scale almost beyond imagining, yet all, it seems, mere ripples in a greater hyperdimensional pond.

Long ago shamen believed that every stone and tree had a guardian spirit, while a modern mystic might say the same of the coiling plasmas that shape nebulae and galaxies. Our place in space may be one of an infinitude, or as unique as the rarest of gems.

Life, and an awareness to experience it with, is blessing beyond compare.

Great Walls and Voids

seriously large-scale structures

Our journey begins at the edge of the observable universe, some 14 billion light-years away. Out here on the perimeter of the known the universe resembles a froth of bubbles.

The bubbles themselves are mind-numbing expanses of virtually nothing, near empty voids that stretch for hundreds of millions of light-years. Surrounding the voids are immense walls, glowing aggregates made up of vast numbers of galaxies.

The walls are of awesome proportions, huge networks interconnected by galactic filaments and sheets that stretch across the firmament. Estimated at 3.5 billion light-years long, by 2.5 billion wide and 50 million thick, the Centaurus Wall (*opposite*) is one of several that form the largest structures so far discovered.

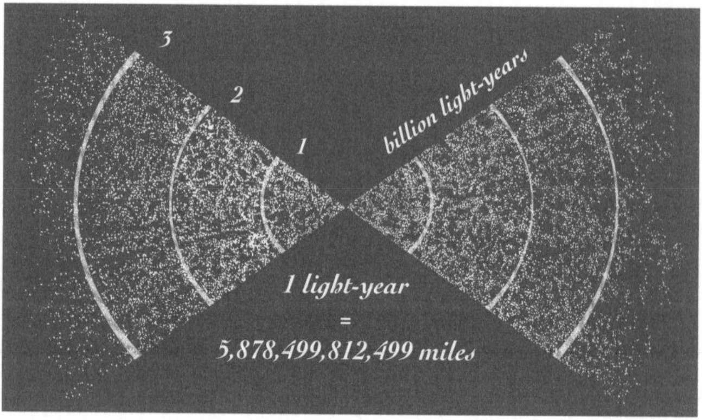
3
2
1
billion light-years
1 light-year
=
5,878,499,812,499 miles

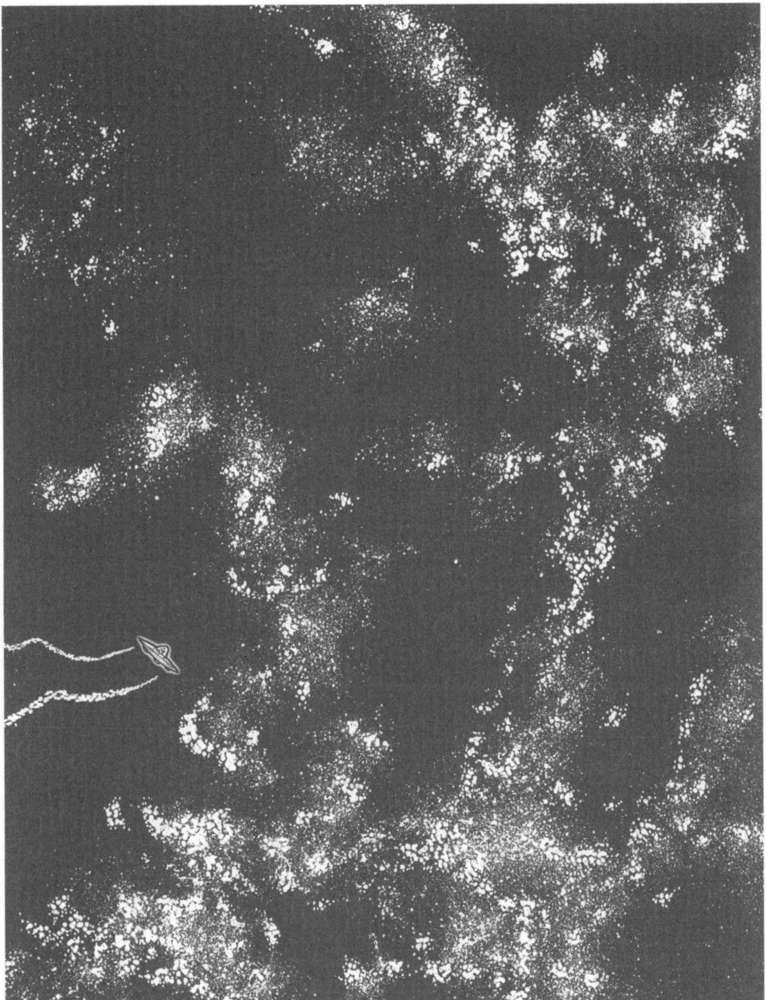

Deep Space

twixt starbursts and galaxies

Surprisingly, the vacuum of space, emptier by far than the best made in laboratories, still has an intriguing population.

Inside galaxies, in the gaps between stars, we find the interstellar medium, a thin mix of gas and dust. Made up mostly of primordial hydrogen and helium left over from the very earliest times, we find here on average one atom per cubic centimeter clumped into cool, filamentary clouds surrounded by sparser but warmer regions.

Other heavier atoms are present in much smaller quantities, drifting leftovers from previous generations of stars. Combining into simple molecules such as ammonia, water, or carbon monoxide, and more complex ones like alcohol, giant molecular clouds can evolve into stellar nurseries. Such clouds may have over a thousand molecules per cubic centimeter, increasing up to a million times more in their denser star-birth regions.

The vast expanses between galaxies also have the odd atom of hydrogen floating about, a few for every cubic meter of space. Even in this intergalactic medium there are sparse traces of heavier elements, roughly one for every million hydrogen atoms. Explosive starburst activity and galactic collisions are believed to have thrown these lonely wanderers out into deepest space.

Though a long way from anything else, these atoms are a torrid bunch, spitting out high energy x-rays with temperatures that can reach several million degrees.

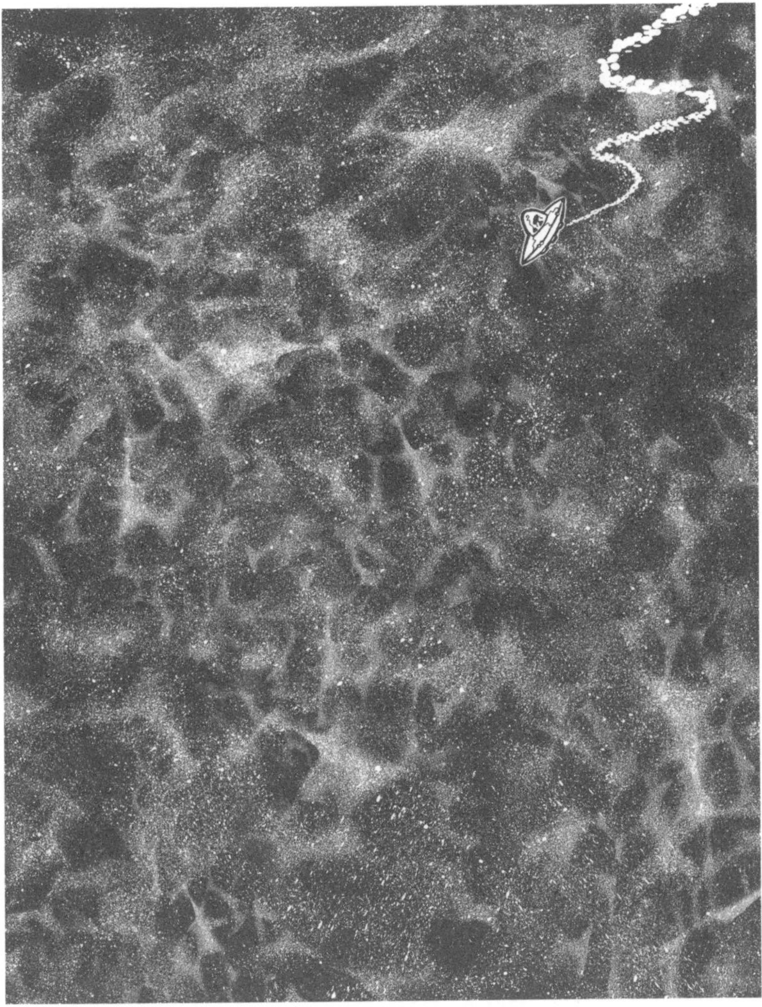

SUPERCLUSTERS
galactic gatherings

The great walls, filaments, and sheets are themselves chains of superclusters, huge aggregates of galaxy clusters linked by thinner skeins of galaxies and all bound together by gravitation. The largest superclusters may stretch for 300 million light-years.

We belong to a modest supercluster 150 million light-years in extent and centered on the Virgo cluster. Our *Virgo supercluster,* along with several others, is heading toward the *Great Attractor*, an enormous mass hidden deep in the Centaurus Wall.

Clusters themselves can be roughly divided into two types: rich clusters, up to 10-30 million light-years wide, which can contain 10,000 galaxies and tend to be dominated by one to three giant elliptical galaxies, and the more common poor clusters, upwards of 3 million light-years across, with as few as ten galaxies, predominantly spirals.

Our *Local Group* is a poor cluster lodged in the outskirts, some 60 million light-years from the Virgo cluster. Orbiting a common center of gravity between ourselves and *Andromeda*, this group of thirty or so galaxies of various sizes covers four million light-years. Three of our neighbors are visible to the naked eye as glowing wisps. The large and small *Magellanic Clouds* are closest, being 170,000 and 190,000 light-years away respectively.

The third is mighty Andromeda, twice the mass of our own *Milky Way*. Currently over 2 million light-years away, the two will merge in about 5 billion years. Although stars rarely bump, gravity's tides will sweep the galaxies into a whirling dance.

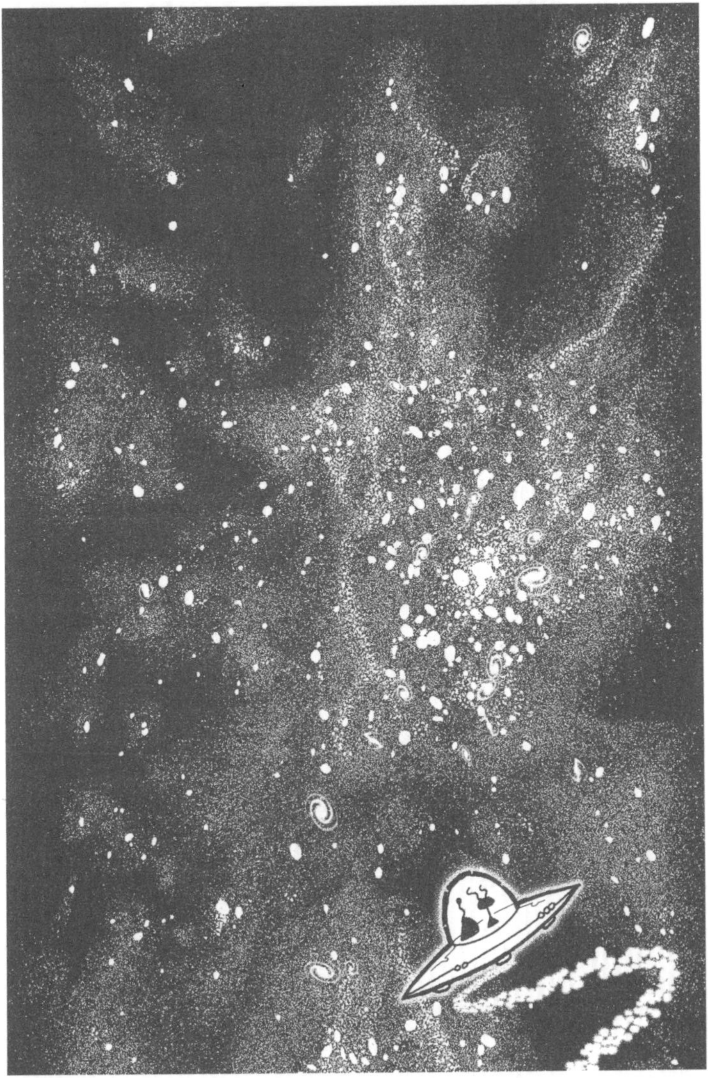

GALAXY TYPES

whorls and wisps

Roughly three-quarters of visible galaxies are, like our own, *spirals,* flattened disks with a central bulge. Young stars, along with gas and dust, hang in the disk. Older stars prefer the bulge or amble the surrounding spherical halo in groups. Spirals come in normal and barred varieties, depending on their centers.

Mainly populated by older stars, the less structured *elliptical* galaxies are roughly spherical with differing degrees of flattening. In between, *lenticular* galaxies are a heady mix of both types.

Harder to fathom are *irregular* galaxies, with shapes sculpted by gravitation or collision (*lower opposite*). Most are star formation hotbeds, with dense clouds of gas amidst crowds of young stars.

Dark and difficult to detect, surveys suggest that dwarf and low surface brightness galaxies perhaps outnumber all the rest.

For ease, galaxies are classified by the Hubble Sequence. Rather optimistically designed to explain galactic evolution, it groups by shape and features (*below*). Our own Milky Way galaxy is type Sb or Sc, but with an additional slight bar on the bulge.

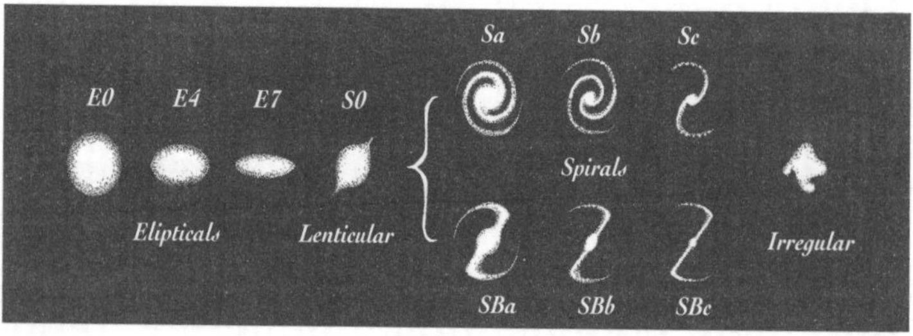
Sa
Sb
Sc
E0
E4
E7
S0
Spirals
Elipticals
Lenticular
Irregular
SBa
SBb
SBc

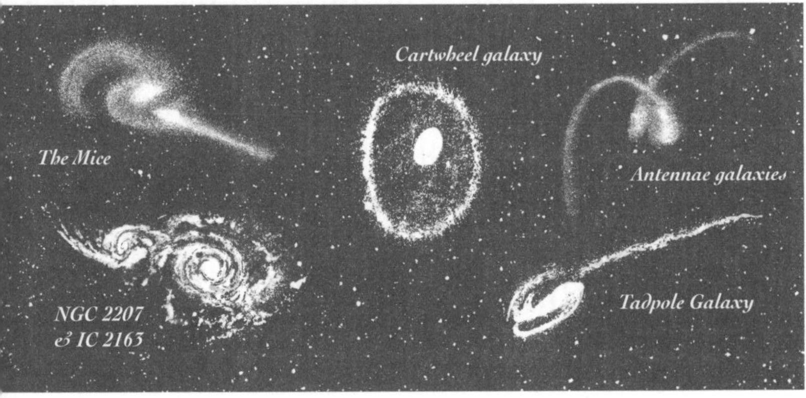
Cartwheel galaxy
The Mice
Antennae galaxies
Tadpole Galaxy
NGC 2207
& IC 2163

ACTIVE GALAXIES
and quasar questions

With ingenious devices scanning the heavens in spectrums beyond the visible, many marvels have been revealed in recent years, and some of the most awesome are the *active galaxies.*

Spewing jets of matter many light-years long from a relatively small nucleus, *radio galaxies* are brightly lit in the x-ray and radio bands even though they may be optically comparatively dim.

Seyfert galaxies are spinning discs with intensely active nuclei. Slightly less bright than Seyferts, elliptical *N-galaxies* have nuclei that vary in intensity. A sub-species, *BL Lacertae* (*BL Lac*) *objects,* have nuclei that are even brighter and more wildly fluctuating. Here it seems we are looking head-on, directly into a jet.

Quasars (*quasistellar radio sources*) are a hundred to a thousand times brighter than normal galaxies and emit vast quantities of radiation (*opposite top*). It is thought that they are the bright hearts of active galaxies, powered by supermassive black holes.

With very high redshifts indeed, quasars are assumed to be incredibly distant young galaxies from a much earlier age of the universe, but a number of quasars, often in pairs with similar redshifts, look like they may have been ejected from other active galaxies (*lower opposite*). They may be recently born galaxies. Objectors point to the quasars' greater redshifts and infer they must be farther away and, so, unrelated or gravitationally lensed.

If the redshifts are anomalous it raises the possibility that our distance estimates may be somewhat inaccurate and that perhaps light could behave differently in high energy zones.

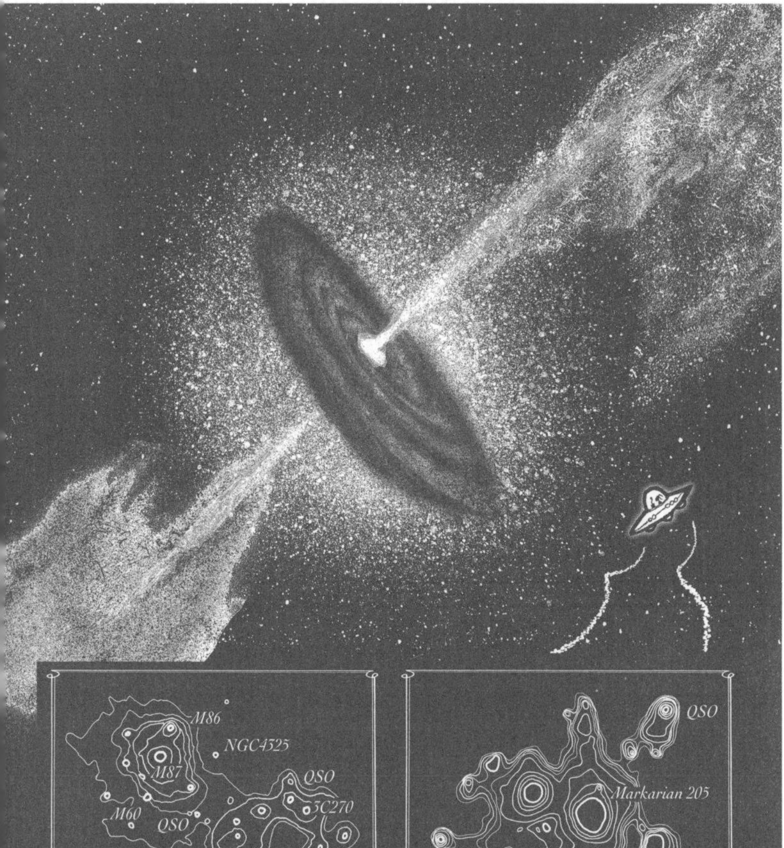
M86
NGC4325
M87
QSO
3C270
M60
QSO
QSO
3C273
QSO
Markarian 205
QSO

X-ray contour maps of the Virgo Cluster (left) and the Seyfert galaxy Markarian 205 (right) showing apparent bridges between supposedly distant high red-shift quasars (QSO) and lower red-shift 'foreground' objects.

Black Holes

through the event horizon

With bizarre physics and gravitational fields so high that light is trapped, black holes are awesome matter vacuums, concentrating enormous amounts of stuff into the tiniest of spaces.

Massive black holes are thought to anchor most galaxies. Millions of times the sun's mass, they may co-evolve with their galaxy, one shaping the other. Supermassive holes, of 10 to 100 billion solar masses, may lurk at the cores of active galaxies and power the fantastic outpourings of quasars.

Stellar-class black holes are lightweights of 3 to 14 times the sun's mass, the collapsed remains of large stars. Some evidence of rarer intermediate-mass holes has also been found.

Those foolhardy enough to approach a black hole (1) would first notice Einstein rings, as light from background stars is lensed by the increasing curvature of spacetime (2). The point of no return is the event horizon, where light is pulled into an endlessly orbiting photosphere and all within is forever hidden (3).

To those outside, the traveler seems to slow to a stop at the horizon's edge, frozen in time, redshifting, and fading (4). The gravity gradient can become so steep that the unwary undergo spaghettification, with head stretched miles from toes (5).

Inside a black hole, time warps into a spatial dimension, space-time compressing to an infinitely dense, inescapable singularity (6). Ripping the fabric of the universe, wormholes may funnel through to other universes. After a massive radiation burst at the worm-hole's mouth, a white hole might come as light relief (7).

fig 1.
fig 2.
fig 4.
fig 3.
fig 5.
fig 6.
Spinning black hole
Non-spinning black hole
fig 7.

The Local Galaxy

meeting the milky way

Approaching from a distance, our own galaxy appears as a starry whirlpool. We first encounter the halo, a spherical shell of dust, gas, and the occasional glob of old stars, that stretches out about 130,000 light-years, deep into intergalactic space.

Elegant spiral arms trace a flattened disk, roughly 100,000 light-years across and 1,000 thick. The arms, bright with young and newborn stars, are thought to result from slowly circling density waves. In between lie dark dust lanes. Our solar system is quietly tucked away in the Orion Arm (*arrowed*) about two-thirds out.

In the middle is the bulge, some 25,000 light-years across and populated with older stars. In the center, anchoring it all, lurks *Sagittarius A*★, assumed to be a spinning black hole of over 4 million solar masses (*below*). Surrounding this are thread-like plasma structures and supernova remnants (*SNR*), all orbiting at a slight tilt to the main galactic disk. A flock of neutron stars and smaller black holes can also be found in the vicinity.

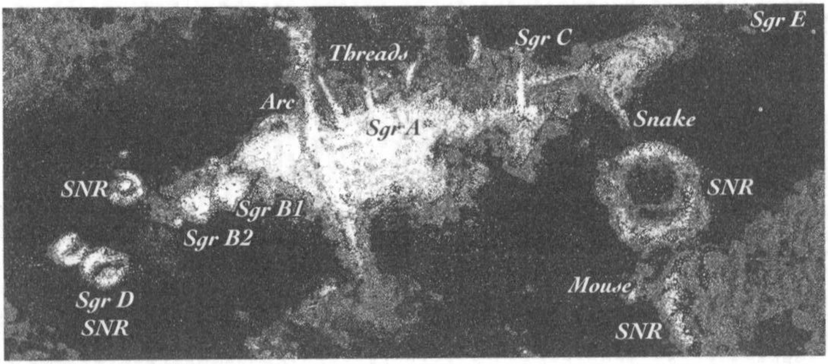
Sgr E
Sgr C
Threads
Arc
Sgr A*
Snake
SNR
Sgr B1
Sgr B2
SNR
Sgr D
SNR
Mouse
SNR

GLOBULAR CLUSTERS

old stars and young stars

The stellar inhabitants of galaxies like ours are divided into two camps, old (*population II*) and younger (*population I*) stars.

Born in the earliest epoch of the galaxy, and estimated to have been around for 15 billion years, the older stars are mainly cool red giants, found either orbiting in the central bulge or grouped together in wandering globular clusters (*opposite top*).

With eccentric paths taking them far from the galactic plane, globular clusters hold tens of thousands to several million stars. Held together by gravity, and possibly centred on medium-sized black holes, these mature congregations measure from sixty to three hundred light-years across. Their leisurely circuits last millions of years, swooping up to 300,000 light-years out into the halo before plunging back through the disk (*lower opposite*).

Meanwhile, the young population I stars are found either alone or mingling in loose open clusters in the gassy, dusty nursery regions of the spiral arms. These stars have a higher heavy element content than their population II cousins, having been seeded from the fusion-factory remains of older stars. With near-circular orbits about the galactic hub, young stars are happy to remain within the 300 light-year thickness of the disk.

An earlier generation of live fast, die young *population III* stars may have existed. Fueled solely on hydrogen and helium, their short lives would have ended in huge supernova explosions that flung freshly forged elements across the youthful galaxy.

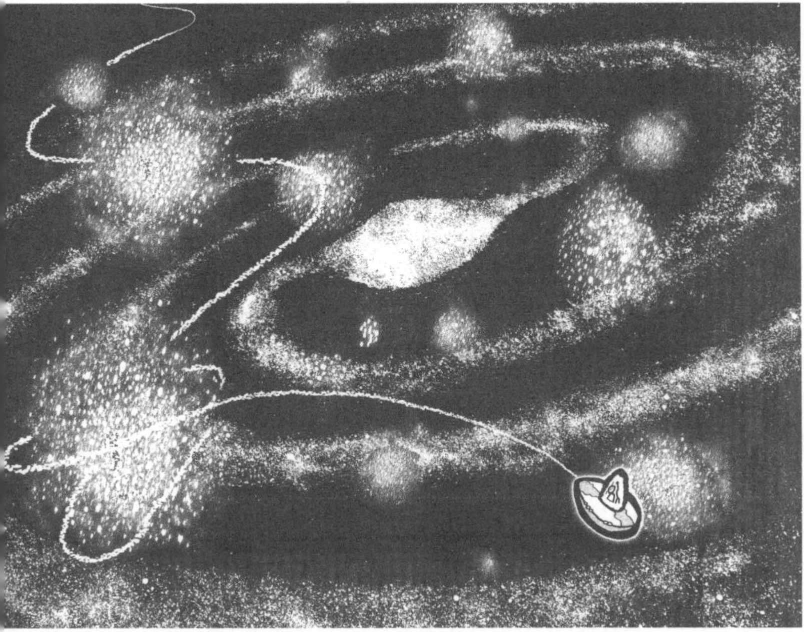

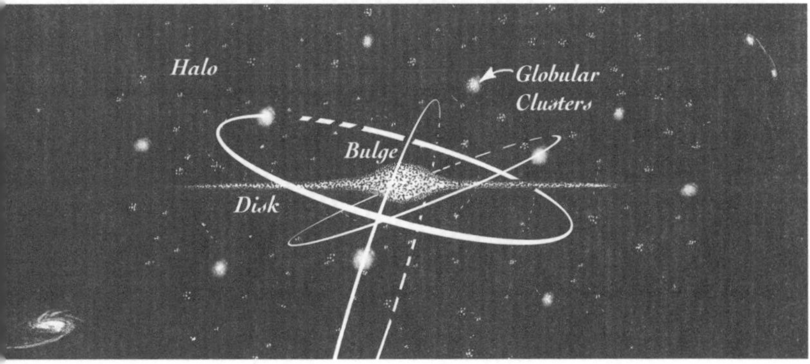
Halo
Globular
Clusters
Bulge
Disk

The Main Sequence

growing up, on and off

From humble cold clouds of interstellar gas, slowly coalescing under gravity into *protostars*, stars evolve in a number of ways depending on mass. Theirs is a balancing act between gravity pulling inward and thermonuclear reactions pushing outward.

For protostars with a third of the sun's mass, gravity squeezes the core to a temperature high enough for nuclear fusion to convert hydrogen into helium. These *red dwarfs* use their fuel frugally, shining for hundreds of billions of years before fizzling out into a *black dwarf*. Protostars of less than a tenth of a solar mass skulk about as *brown dwarfs,* where fusion can't even start.

Mid-mass stars like the sun take a different path. When their fuel is used, the core collapses. Atoms squish together, boosting temperatures. The outer hydrogen shell ignites, puffing out into a *red giant*, and the by now helium core starts fusing into carbon. Finally the helium is gone and the core collapses again, but lacking enough mass to generate sufficient heat to further convert the carbon the star fades into a hot, compact *white dwarf*.

Stars of three solar masses or more lead fast, furious lives, rapidly passing through the early stages. Depending on mass, carbon goes on to fuse into neon, then oxygen, silicon, and finally iron when fusion stalls, requiring energy input to continue. During these processes the star swells into a *supergiant,* hundreds of times the sun's diameter. Alas when fusion stops, gravity takes over and the core collapses, culminating in a vast, iron-rich explosive *supernova,* which also forges many other heavy elements.

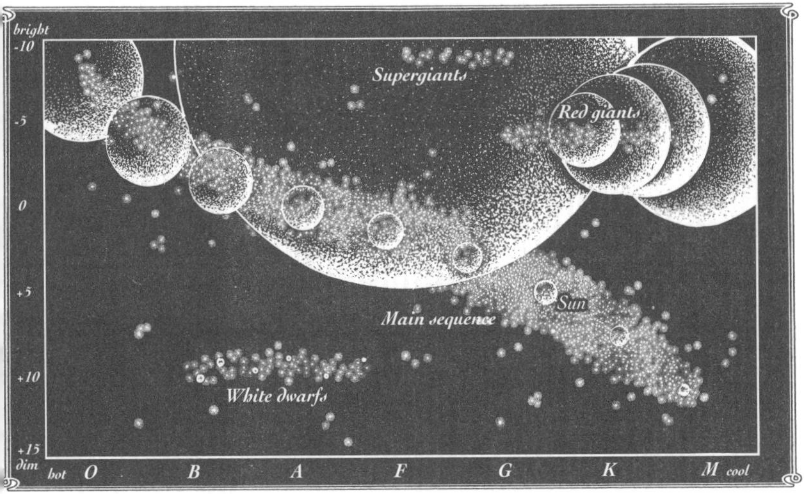
bright
-10
-5
0
+5
+10
+15
dim
hot O
B
A
F
G
K
M cool
Supergiants
Red giants
Sun
Main sequence
White dwarfs

The Hertzsprung-Russell diagram plots stars according to their luminosity, expressed as absolute magnitude (vertical scale), and spectral class (horizontal scale). Running diagonally across the diagram is the main sequence where most stars spend the greater part of their lives.

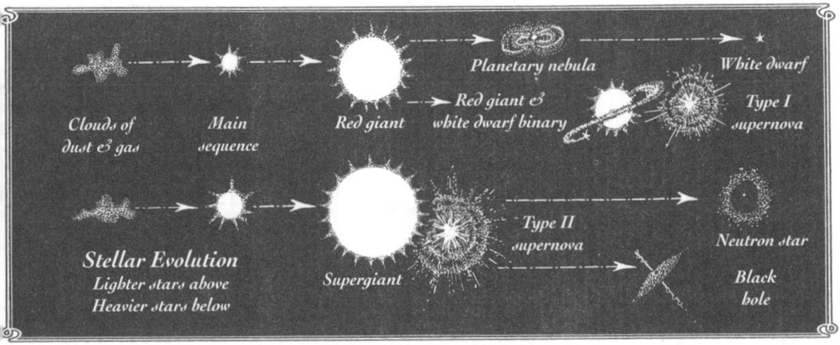
Planetary nebula
White dwarf
Red giant & white dwarf binary
Type I supernova
Clouds of dust & gas
Main sequence
Red giant
Type II supernova
Neutron star
Black hole
Supergiant
Stellar Evolution
Lighter stars above
Heavier stars below

The eventual fate of a star depends on its mass, the larger the star the more spectacular its demise.

Nebulae and Supernovae

stellar birth and death

The beginning and end of a star's life offer some of the universe's most spectacular sights. Consisting primarily of vast clouds of hydrogen, nebulae come in a number of guises.

Dense regions of interstellar gas and dust are the delivery rooms for *stellar nurseries*, aglow with the light of new stars.

Young hot stars excite the blushing plasma of *emission nebulae*, their ultraviolet light stripping electrons from hydrogen gas, later glowing red when gas and electrons eventually recombine.

Dusty clouds bathed in the light of nearby stars cause the often blue *reflection nebulae*, the Pleiades hosting a great example.

The end of a red giant or supergiant's life arrives when the outer layers of the star are blown off to create *planetary nebulae*, shells of plasma and dust about the exposed core (*opposite top*).

Massive stars and some types of binary star systems end in collosal supernova explosions. Shooting outward, heavy elements synthesized in the star's heart are strewn across space ready to make the next generation of worlds, while the shockwaves may jostle interstellar gas into birthing new stars.

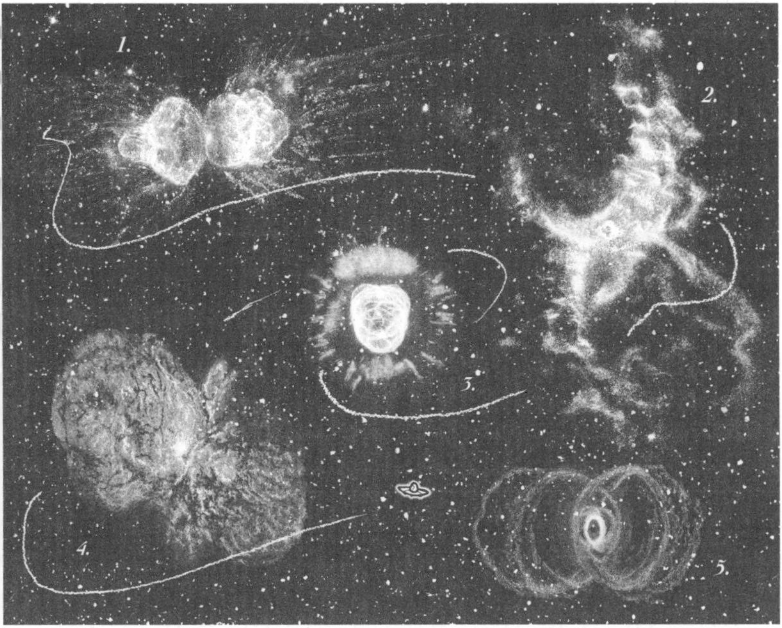
1.
2.
3.
4.
5.

Named because of the mistaken idea that the remaining core was a planet, planetary nebulae are one of nature's most beautiful pieces of handiwork. Above are some picturesque examples: (1) Ant nebula Mz 3, (2) Red Spider nebula, (3) Eskimo nebula, (4) Eta Carina, (5) Hourglass. The Horsehead nebula (left) is part of a large complex of molecular clouds, emission nebulae and star nurseries that form the Great Nebula below Orion's belt. The light-echo sequence (below) developed over a period of several years as a burst of light from the star V838 Monocerotis passed through a surrounding dust cloud.

Neutrons and Novae

lighthouses and candles

The dazzling supernova deaths of ample four to eight solar-mass stars create exotic iron-shelled *neutron stars.* Gravity, no longer balanced by fusion, crushes the remains of the core. Atoms are scrunched so tightly that all the electrons within combine with protons, leaving nothing but neutrons.

Though rather compact, a mere ten miles or so across, neutron stars are incredibly dense: a thimbleful on Earth would weigh as much as a mountain. Such high densities lead to hugely strong gravitational and magnetic fields.

Spinning neutron stars create jets of energized matter that emit intense beams of light and radio waves. When a beam sweeps the Earth, the star pulsates like a celestial lighthouse. The fastest *pulsars* blink nearly a thousand times a second.

Useful for astronomers, *Cephid variables* are stars whose size and luminosity vary with clockwork regularity. Cephids with the same period are equally as bright, so one appearing dimmer has to be farther away. This means they can be employed as standard candles, enabling the distance to other stars in their vicinity to be estimated.

Our sun, being a singleton, is in the minority. Most stars have companions in binary or triple systems, or live in even larger clusters. In a binary pair where a small hot star, often a white dwarf, and a large star orbit, the smaller star can drag streams of matter from its partner. This gas builds up, gets hotter, and periodically detonates as a blazing *nova* outburst.

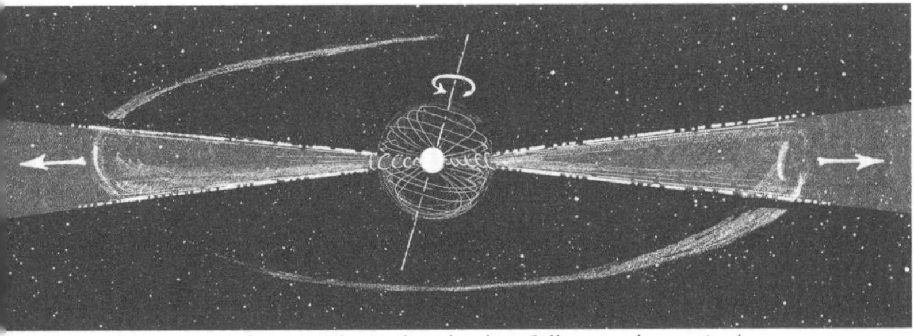

The magnetic and rotational axes of a pulsar lie at different angles, causing beams of highly energized matter to sweep across the universe as the star spins.

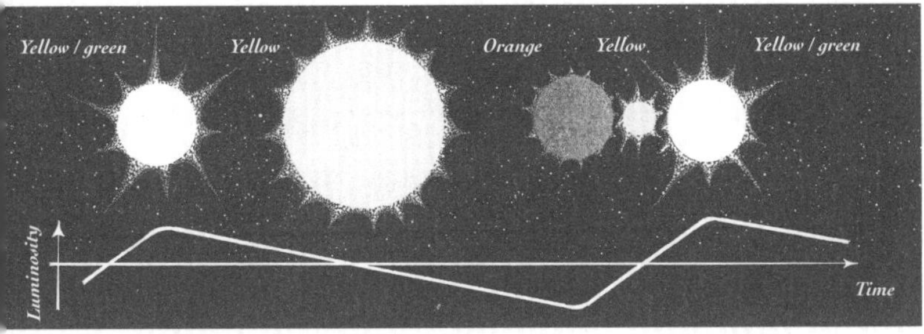

Yellow / green
Yellow
Orange
Yellow
Yellow / green
Luminosity
Time

As a Cephid variable star swells and shrinks, its brightness and color changes.

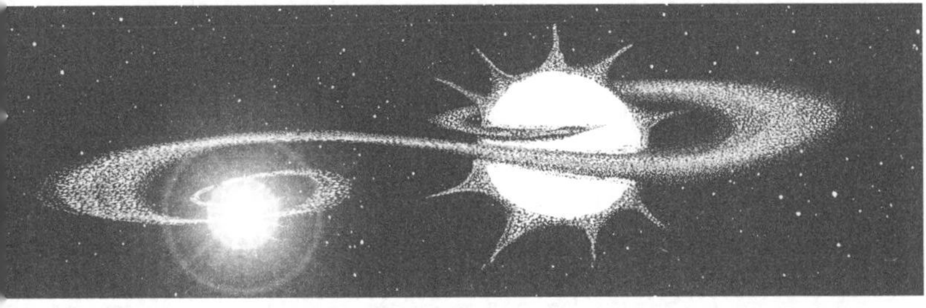

A binary star system

AURIC OBJECTS
to the heliopause and beyond

Stretching out some 4.5 trillion miles, halfway to our closest stellar neighbor, *Proxima Centuri*, the conjectured Oort Cloud is a halo of icy debris remaining from our solar system's birth. Flung into highly eccentric orbits, these distant, dirty snowballs may pay a rare flying visit to the sun as long period comets.

More regular visitors, short period comets fall in from the Kuiper Belt, a loose disk of ice and planetoids extending out some 5 billion miles, twice as far as Neptune's orbit.

Farther in are the planets; Pluto and gas giants Neptune, Uranus, Saturn, and Jupiter, with occasional moons, wayfarers like the Trojans and other asteroids, leftovers from the relentless churnings of gravity. Harmonically orbiting within are the rocky inner planets, Mars, Earth, Venus, and Mercury.

From the center comes the solar wind, a rush of electrically charged particles blowing into space forming the heliosphere, a huge magnetized bubble of plasma around the Sun. Slowing to below the speed of sound at the termination shock about 10 billion miles out, the ions eventually stream into the galaxy's realm at the heliopause some 14 billion miles away.

The last ripples of the sun's magnetic grip fade around 21 billion miles out, the bow shock of the solar system as it ploughs its way through the interstellar medium.

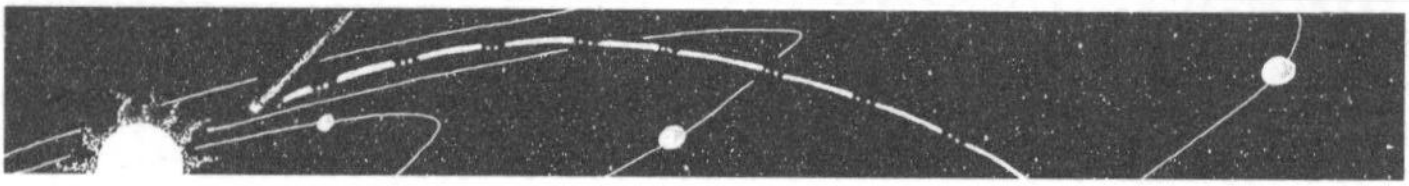

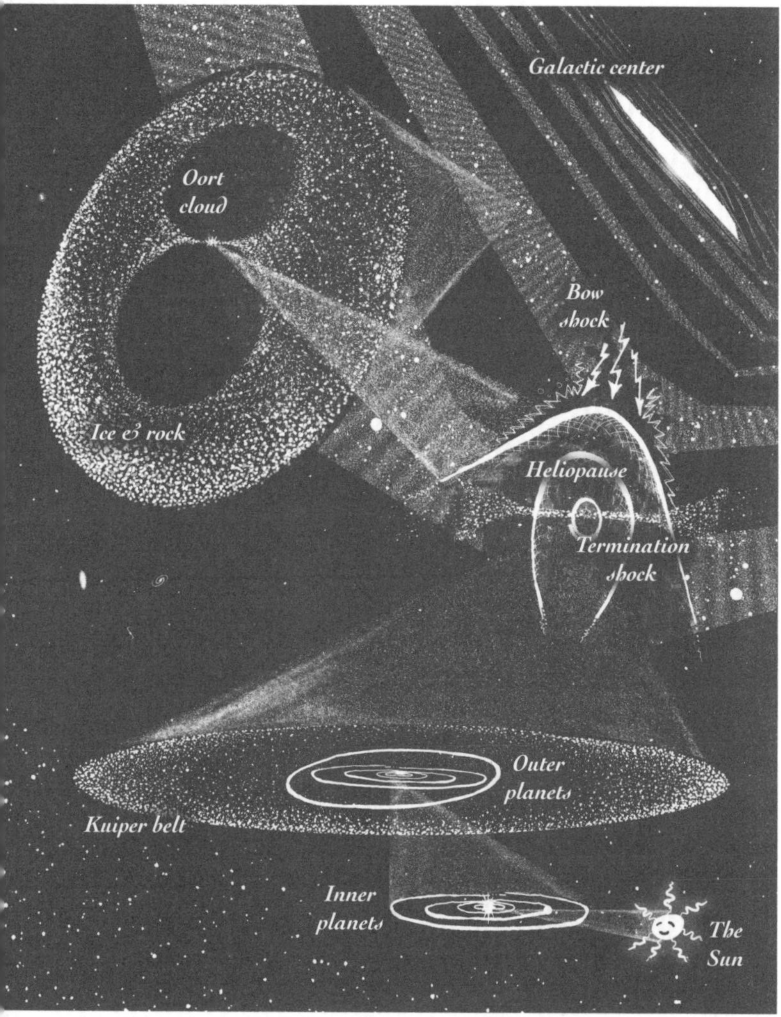

Galactic center
Oort
cloud
Bow
shock
Ice & rock
Heliopause
Termination
shock
Outer
planets
Kuiper belt
Inner
planets
The
Sun

PLANETS AND MOONS

forming solar systems

According to current theory, some 4.6 billion years ago a huge cloud of dust and gas grew lumpy and collapsed under gravity, spinning and flattening into a disk. At the center the largest lump became so massive that hydrogen fused into helium, igniting the nuclear fires of our sun. Smaller planetessimals, orbiting in the surrounding accretion disk, coagulated into a collection of planets (*opposite top*). Like a giant mixer on full speed, some parts ferociously collided while others were flung out into the Oort Cloud or farther into deep space. All this may have occurred in a mere 100,000 years, though recent models suggest that it could have happened even quicker - in the wink of a cosmic eye.

Our own solar system (*below*) is just one of many. Hundreds of exoplanets in extraordinary orbital systems have been detected from the slight wobblings of nearby stars (*lower opposite*).

Many have masses greater than our largest planet, Jupiter, and may be brown dwarfs, protostars not quite big enough to burst into nuclear fusion. Some of these giants have orbits of days rather than months, whipping around extremely close to their parent star. Quietly behaved systems like ours seem quite rare.

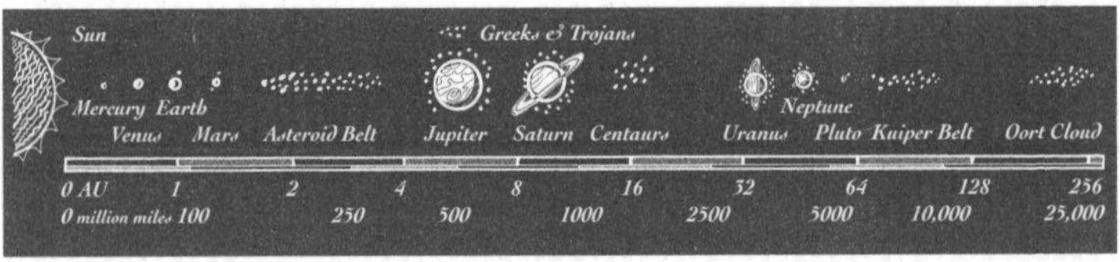
Sun
Greeks & Trojans
Mercury
Earth
Venus
Mars
Asteroid Belt
Jupiter
Saturn
Centaurs
Neptune
Uranus
Pluto
Kuiper Belt
Oort Cloud
0 AU
1
2
4
8
16
32
64
128
256
0 million miles
100
250
500
1000
2500
5000
10,000
25,000

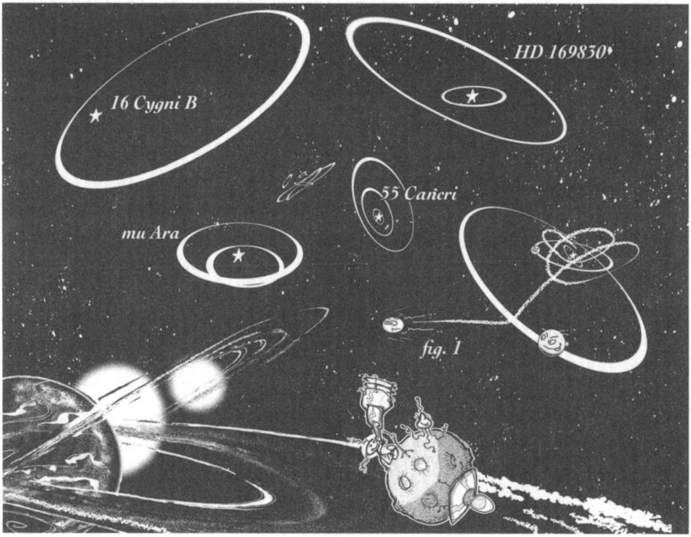
HD 169830
16 Cygni B
55 Cancri
mu Ara
fig. 1

The Neighborhood Star

a glimpse of the sun

Providing most of life's energy on this planet, our sun is a colossal nuclear fusion reactor that forges two hydrogen atoms into one helium while giving off heat, light, and other radiation as by-products. It can take hundreds of thousands of years for a photon of light to barge its way out of the seething interior, and then a fleeting eight minutes to travel the 93 million miles here.

At about 4.6 billion years old, our sun is nearly halfway through its life. Its present diameter of 870,000 miles will eventually bloat to roughly the Earth's orbit, frying us all before shrinking and retiring as a white dwarf, a cinder the size of the Earth.

The sun is mostly superheated plasma, gas so energized that electrons are stripped from their atomic nuclei. Thousand mile wide convection bubbles boil across the surface while intense twisting magnetic fields create majestic loops and prominences.

The sun's coiling magnetic field reverses every twenty-two years, and *sunspots,* planet-sized windows into a marginally cooler inner layer, appear during each half of the cycle (*below*).

Periodically, billions of tons of plasma burst forth in titanic coronal mass ejections, roiling clouds some 10 million miles wide that move at up to 5 million miles an hour. Solar weather can disrupt electronics, cause power surges, and produce spectacular polar aurora displays around our little world.

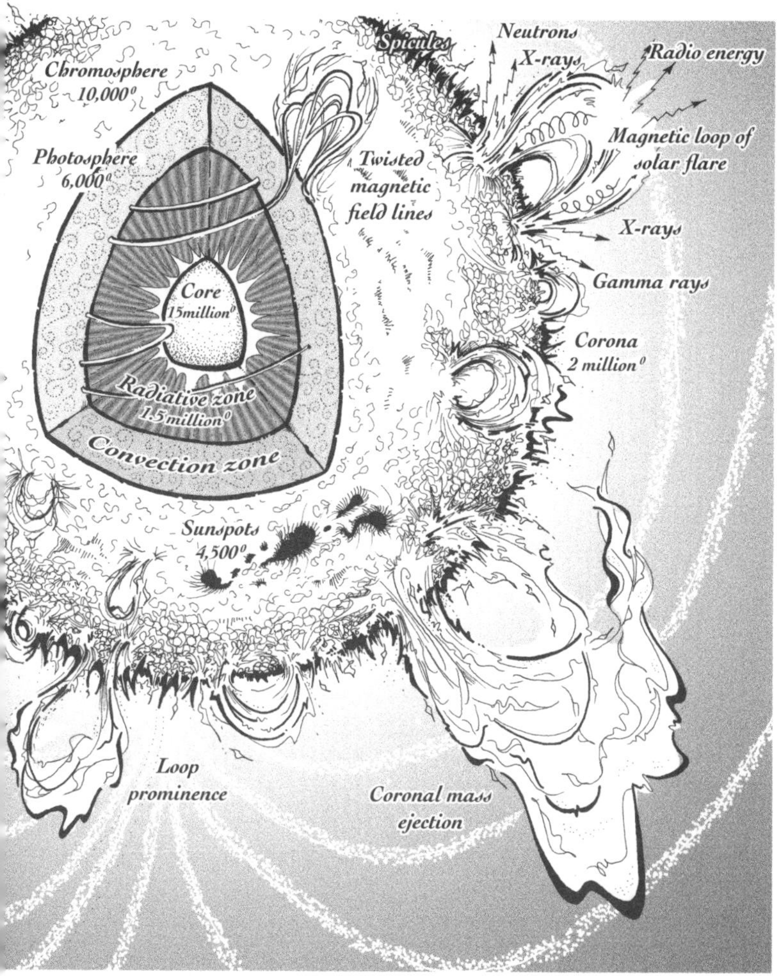
Spicules
Neutrons
X-rays
Radio energy
Chromosphere
10,000°
Photosphere
6,000°
Twisted
magnetic
field lines
Magnetic loop of
solar flare
X-rays
Gamma rays
Core
15million°
Corona
2 million °
Radiative zone
1.5 million°
Convection zone
Sunspots
4,500°
Loop
prominence
Coronal mass
ejection

Composition of the sun : 70% hydrogen, 28% helium, 2% heavier elements

The Radiation Cosmos

hidden skies for electronic eyes

Our sense of sight recognizes a narrow band of electromagnetic frequencies we call visible light. Today we are able to glimpse beyond, our electronic gaze revealing incredible wonders.

Radio waves have the longest wavelength. Bright in this band are the spectacular jets that stretch for thousands of light-years from the fervid cores of active galaxies. Vast, glowing molecular clouds swathe the sky, while the radiant remains of supernovae allow peeks into the maelstrom of our galaxy's heart.

Raising the vibration a notch, everything seems bathed in a near uniform cosmic *microwave* background radiation. Perhaps a relic of the earliest epochs of the universe, small ripples may hold clues as to how today's large structures evolved. Syncopated microwaves also splash from maser fountains around new stars.

At higher frequencies the stars begin to shine in the *infrared*; from the youngest, coddled in warm, dusty nebulae, through bright starburst galaxies, to ageing white dwarfs and giants.

On the other side of the visible lies the *ultraviolet*. In this realm hot stars and supergiants shine brightest and young UV galaxies blaze with newly born hot blue stars.

Tenuous galactic plasma shimmers in the shortwave *x-ray* bands, while dazzling outbursts mark extreme events, possibly surrounding the gravitational maws of ferocious black holes.

At greater energies still, the occasional blinding detonation of a *gamma ray* burster flashes, possibly the cataclysmic final roars of supermassive stars in very distant galaxies.

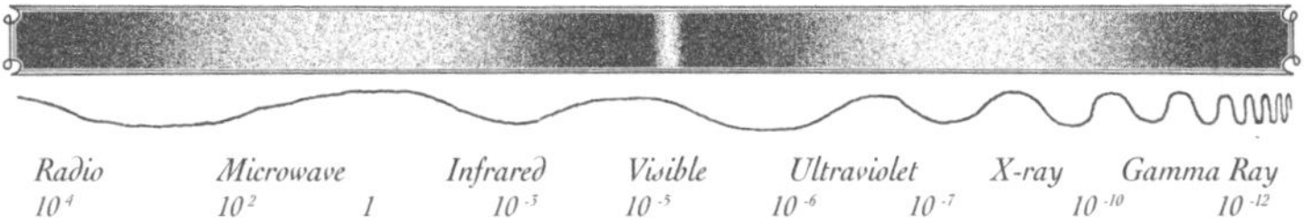
Radio 10^{4}
Microwave 10^{2} 1
Infrared 10^{-3}
Visible 10^{-5}
Ultraviolet 10^{-6} 10^{-7}
X-ray 10^{-10}
Gamma Ray 10^{-12}

The Electromagnetic Spectrum, wavelengths in meters

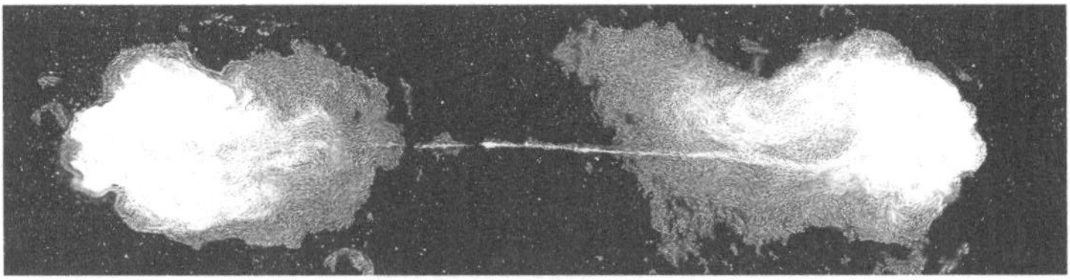

The jets of radio galaxy Cygnus A stretch 18,000 light-years in each direction.

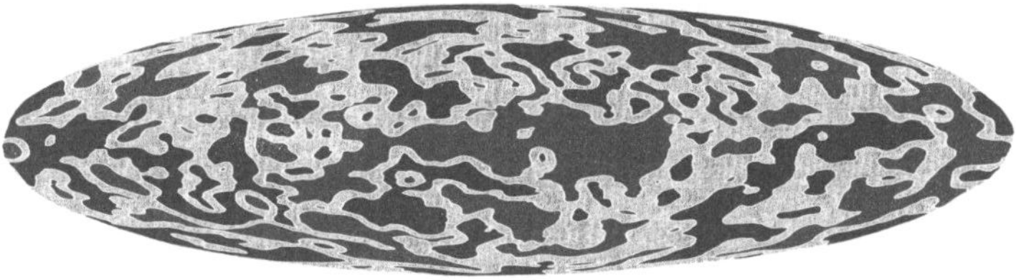

Sky map of variations in the cosmic microwave background radiation. These small fluctuations are believed to have formed the kernels of the later large-scale structure of the universe.

IRc2
Orion
Nebula
Young, hot star
Powerful water masers
Toroidal shape, similar in form to the $3d^{3}$ electron orbital
hydroxyl molecule
water molecule

Incredibly strong masers (microwave amplification by stimulated emission of radiation) form around young stars shrouded by molecular clouds. Water and hydroxyl molecules enter aexcited energy states. Dropping back they emit a buzz that fires up the whole system.

SHIFTING SPECTRA

expanding space

One puzzle facing astronomers has been to find the composition of distant objects. Spectroscopy uses the unique fingerprint frequencies at which atoms and molecules emit or absorb light. Splitting the light received from an object into a spectrum reveals these frequencies as spectral lines, which can then be analyzed to determine the chemical ingredients.

Light can also be used to estimate speed. Approaching objects appear blue-shifted as light waves bunch up toward the blue end of the spectrum. Conversely as an object recedes the light waves are stretched, to become red-shifted. This *Döppler effect* is used to judge velocity by measuring how far the distinctive patterns of spectral lines have been moved along the spectrum.

Observations show that faint, distant galaxies become increasingly red-shifted, suggesting that the farther away an object is, the faster it is racing away from us. The implication of this cosmological redshift is that the whole universe is expanding.

Hubble's Law charts a relationship between a galaxy's brightness and redshift, indicating the immense distances involved. The highest redshifts found have belonged to quasars billions of light-years away, suggesting they are either hurtling off at a significant fraction of light speed or are closer and behaving very weirdly.

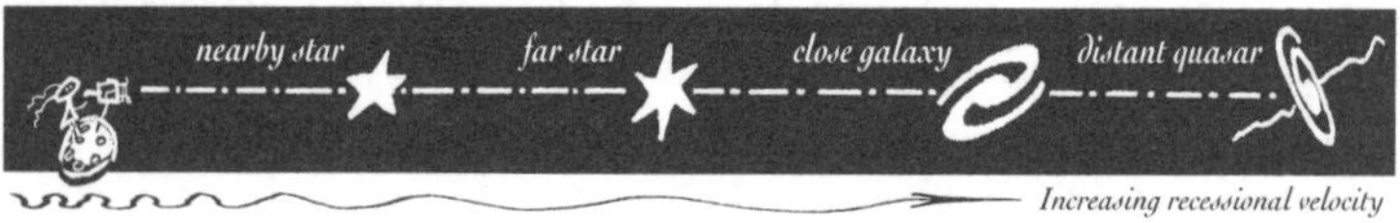
nearby star
far star
close galaxy
distant quasar
Increasing recessional velocity

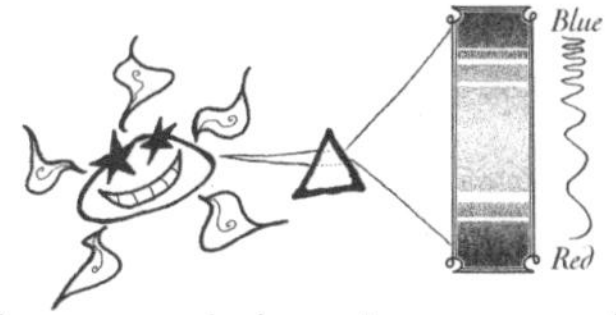
Blue
Red

The spectrum of a hot main sequence star has mainly hydrogen & helium absorbtion lines.

Cooler red giants have a more complex spectrum showing that heavier elements are present.

Blueshift
Redshift

Doppler shifts: an approaching object appears blue-shifted while a receding one will seem red-shifted.

Increasing redshift

Cosmological redshift indicates that the universe is expanding. Like dots painted on a balloon (or alien) which is then inflated, distant points seem to race away faster than closer ones. Observations show the spectra of distant objects have increasing redshifts.

Forces and Fizzicks

known and knot

Holding the universe together are four forces. Two, the *strong* and *weak,* govern subatomic interactions, while electric sparks, attractive magnets, and rays of light are due to the *electromagnetic.*

Although the weakest force, *gravity* extends its grip across the cosmos working its magic by warping spacetime; the greater the mass, the larger the warp. Both space and time are affected, clocks running slightly slower on Earth than in orbit.

Gargantuan events like supernovae and colliding black holes may create huge subtle gravity waves, moving distortions of space-time itself. More visible are gravitational lenses where light from a distant source is bent when passing a massive intervening object to produce spectacular multiple images of the source.

There is a suspicion that the forces are all aspects of one phenomenon. M-theory suggests the fundamental particles of the universe might be smaller strands of vibrating energy, tying string-theories with relativity and quantum concepts. Like notes on a violin, each particle is a particular resonance of the string.

One dimensional strings, and their extended membrane-like cousins the branes, need ten or eleven dimensions in which to vibrate, more than the usual four of space-time. These may exist folded into spaces smaller than atoms, or live invisibly alongside ours. Some scientists believe our universe is confined to a particular brane and that we live on but one of an infinitude of bubbly branes, floating through a multidimensional hyperspace.

The strong force binds together the quarks that make up atomic nuclei.

N
W

The weak force governs radioactive decay & neutrino interactions.

Electromagnetism is not just light & radio waves, but also enables chemistry to take place.

Colliding black holes creating gravity waves that ripple through space-time.

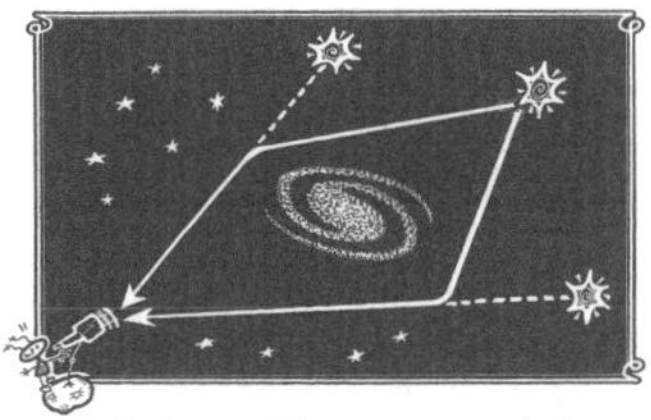

Light from a distant quasar is bent while passing a massive galaxy, creating multiple images of the source.

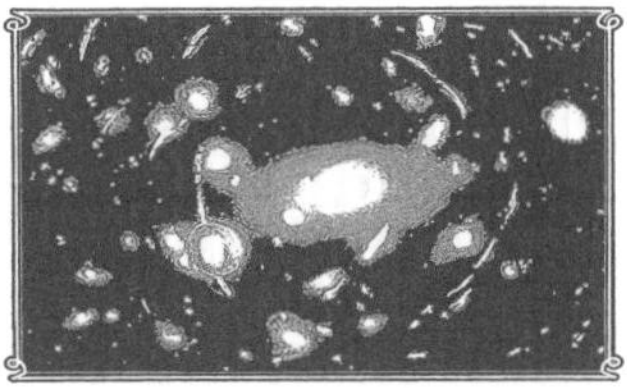

A fine gravitational lens centered on Abell 2218, revealing background galaxies.

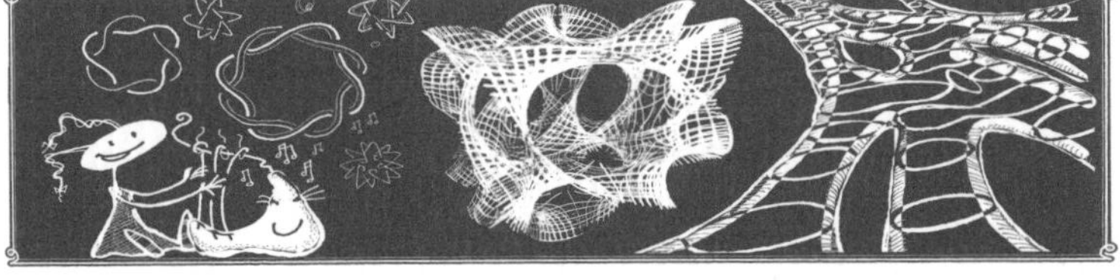

Superstrings, vibrating energy strands, may be the essence of all matter.

BIG BANG OR STEADY STATE?

fabricating a universe

The *big bang* theory is our most popular scientific creation myth, interpreting many, though not all, of the mysteries that surround us. It suggests that some fourteen billion years ago the entire universe burst into being as a searingly hot, infinitesimally small point. Space-time curved into existence from this singularity, held together by one unified force. As the new universe rapidly cooled, the force crystallized into the four we are familiar with today. Suddenly undergoing inflation it grew enormously, stretching tiny fluctuations in its fabric into seeds for large-scale structures. Matter condensed out from energy, clumping first into sub-atomic particles, then atoms, stars, galaxies, clusters, and eventually us.

Another take is *steady state cosmology.* This leans toward continuous creation in a universe that has always existed. Here hydrogen forms in interstellar space, or possibly galactic cores, slowly collecting into molecular clouds and higher order objects. A *quasi steady state* variant pictures the universe as expanding and contracting over immense periods of time, each phase building on structures remaining from the previous one.

Cosmic in- and out- breath is echoed in the *cyclic braneworlds* model in which our universe coexists alongside another. A pair of brane universes periodically crash together, bouncing off each other to initiate a new round of creation. After trillions of years the initial impetus fades and the universes once more approach, ending one epoch and beginning another with a big splat.

Like everything else, theories are born, reproduce, and evolve.

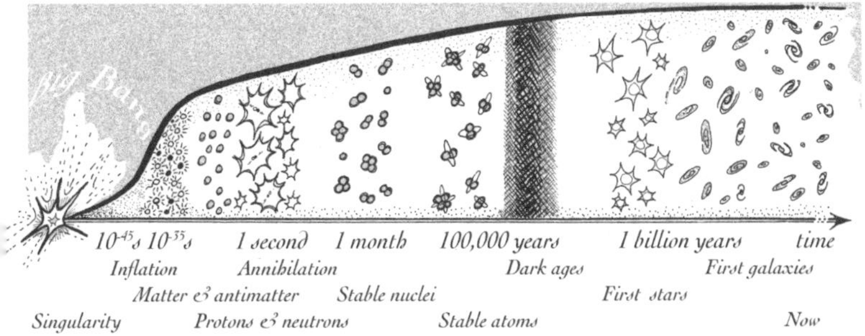
Big Bang
10^{-45}s 10^{-35}s
1 second
1 month
100,000 years
1 billion years
time
Inflation
Annihilation
Dark ages
First galaxies
Matter & antimatter
Stable nuclei
First stars
Singularity
Protons & neutrons
Stable atoms
Now

A quick look at the Big Bang

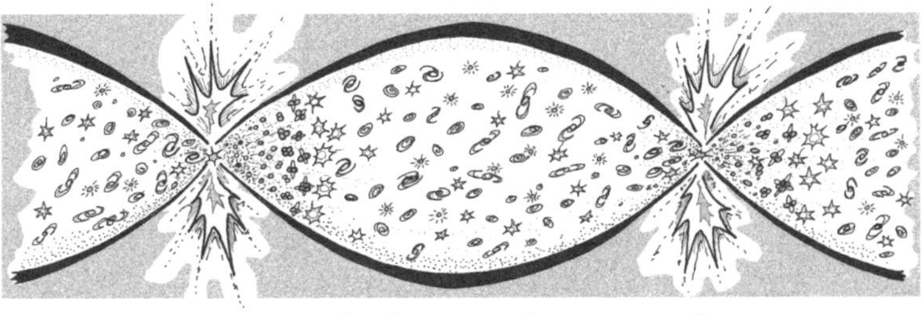

Repeating cycles of Quasi-Steady State Cosmology

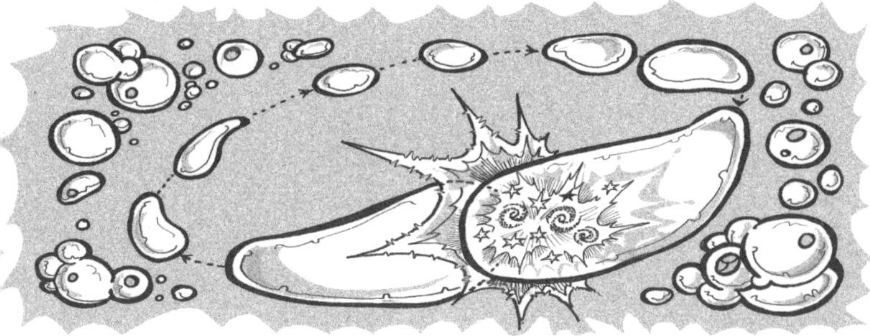

Colliding hyperdimensional branes

Orbital Evolution
and the shape of things to come

Ancient stargazers charted the wanderings of sky-gods across the dome of the firmament. The Earth was at the center of all creation, with celestial spheres encircling the heavens.

The Greeks invented epicycles (*top left*) to model the dance of the planets. Generations of ponderers refined the cranks and wheels until it was audaciously suggested that the Earth spun about the sun (*top center*). Suddenly planetary motions became much simpler.

Nowadays there are many centers, as small things go round bigger ones, all snuggly wrapped in gravity's web and orbiting and dancing around changing common centers of mass (*top right*).

Ultimately the fate of the universe rests on the critical density, a balancing act between gravity, holding stuff together, and inflation, pushing things apart. Several possibilities result.

An *open universe* has too little mass for gravity to stop expansion. Like a three-dimensional saddle curving out forever, matter dilutes over time and entropy slowly takes its toll, leaving little else but gigantic black holes and fading photons.

Where there is more than enough mass to stop expansion, space curls in on itself, resulting in a *closed universe* which achieves a maximum size and then shrinks, ending its days in a big crunch.

Should gravity exactly balance inflation, expansion slows down and a *flat universe* eventually reaches a steady state. Intriguingly, some theoretical finite flat universes have no edges and wrap back on themselves. Even flying in a perfectly straight line any brave travelers would eventually return home.

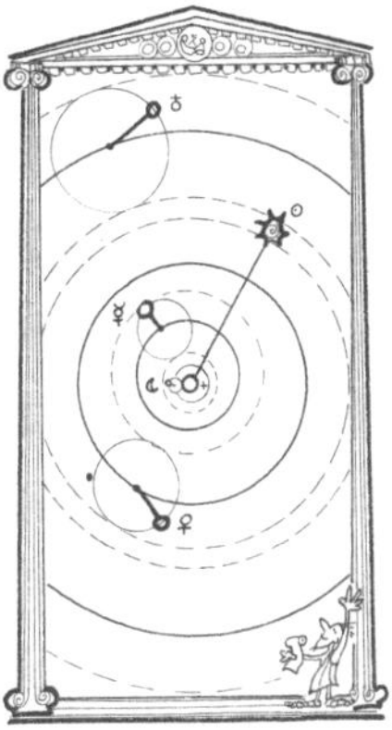

Earth-centered epicycles

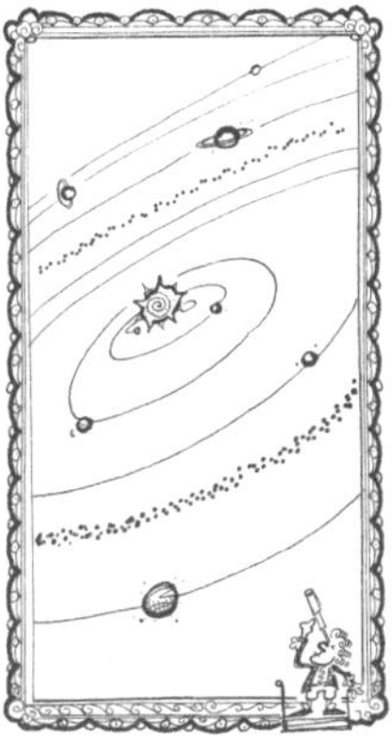

Sun-centered Keplerian orbits

Galaxies revolving around a gravitational center of mass

Open universe

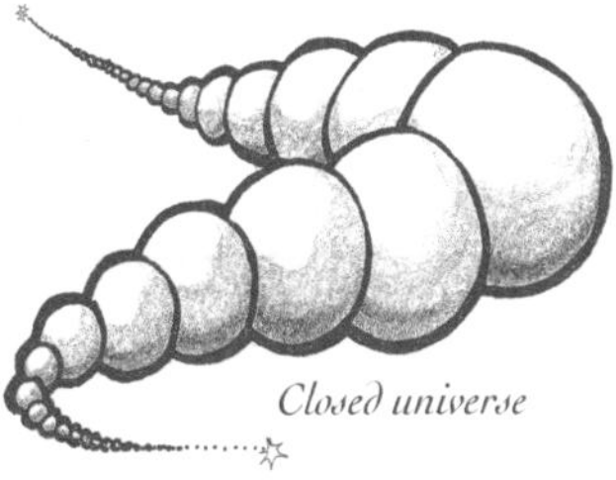

Closed universe

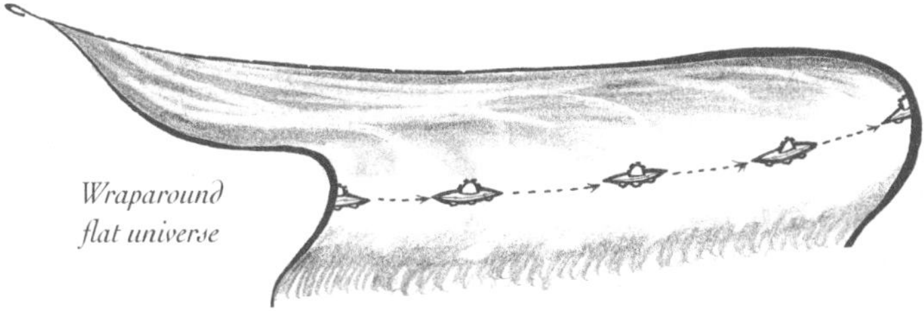

Wraparound flat universe

Cosmic Mysteries

missing matter and other conundra

It is to be expected that a universe as diverse as ours may have the odd surprise hidden up its many light-years long sleeve.

One question is why the universe is mainly matter rather than mirror-image antimatter, as theoretically both are created equally. Nature has a slight bias, which is difficult to explain.

Another problem concerns galactic evolution. Primordial gas collapsing under gravity provides a starting point, but does not reveal all the intricacies of the many and varied galaxy types.

Fans of the big bang scenario had a shock when analysis of globular clusters showed some venerable stars to be 18 billion years old, seemingly older than the entire universe; however this seems small-fry compared to the following headscratchers.

Investigations of the motions of stars and galaxies have shown that for the structures to remain glued together, more mass is needed. Apparently a hefty chunk of the universe is invisible.

To keep the models on track, theorists invented *dark matter*, suggesting all manner of dense and exotic particles, massive neutrinos, or miniature black holes. Woven dark matter filaments have been proposed as seeds for the large-scale bubbly cosmic fabric (*opposite top*), with galaxies forming on the nodes.

Most puzzling of all is that observations of distant supernovae seem to indicate the expansion of the universe is accelerating. Space itself is being pushed apart by a mysterious *dark energy*. If true, we face the strange prospect of having very little clue as to what the majority of our universe is actually made of.

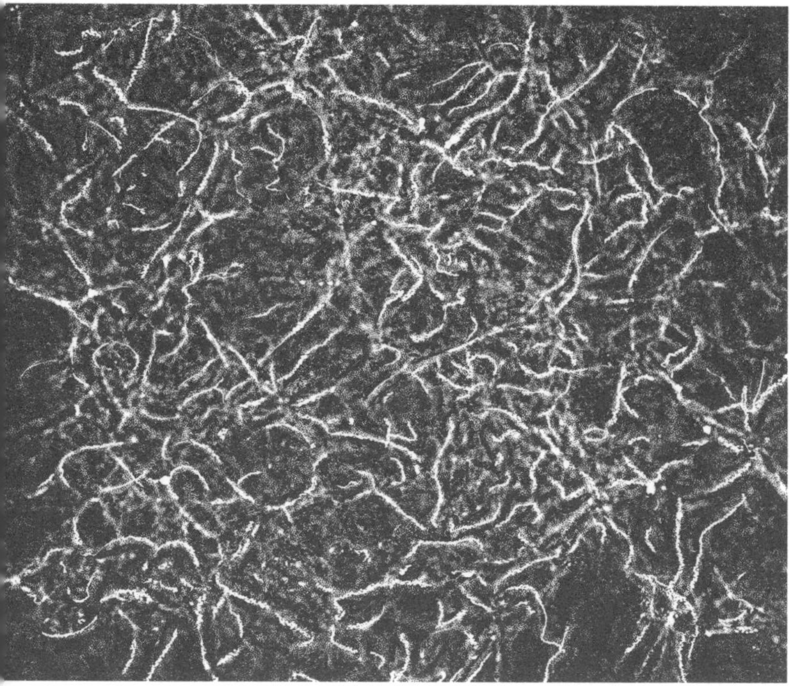

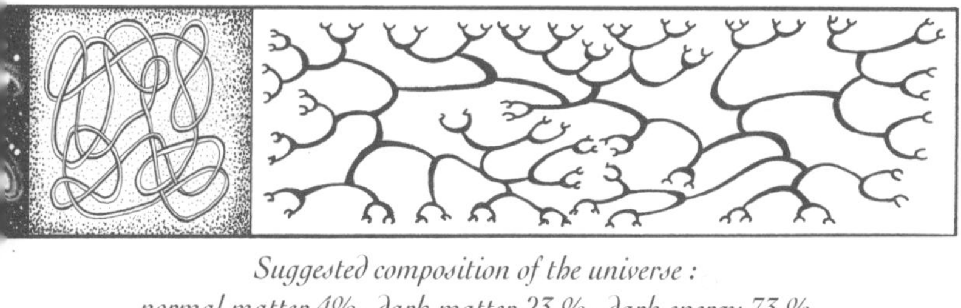

Suggested composition of the universe :
normal matter 4% , dark matter 23 % , dark energy 73 %

A Plasma Universe

into the vortex

A great deal of the visible universe is in the form of plasma, a gaseous, high-energy state of matter where electrons, stripped from their atoms, zip freely about. Plasmas conduct electricity excellently, tightly spiraling electrons through magnetic fields to produce dynamic Birkeland currents that coil into filaments.

Such currents flow between Jupiter and its moons, and filaments over a hundred light-years long, yet only three wide, have been detected in our galactic center. Plasma ribbons reaching far across deep space have been discovered, along with braided jets that shoot at near light speed from the cores of active galaxies. It has even been suggested that the galaxies themselves are acting as giant electrical generators (*opposite*).

Experiments show that at extremely high temperatures plasmas behave like liquids. It is possible that early on in its life the incredibly hot, young cosmos could have been filled by a searing plasma sea of strange subatomic quark molecules.

The maximum size that filaments can achieve before becoming unstable has been calculated to be similar to the largest structures so far observed, the Great Walls. Galactic evolution has also been modelled by twisting currents in the laboratory *(below)*. It seems that plasmas may well hold the keys to many riddles.

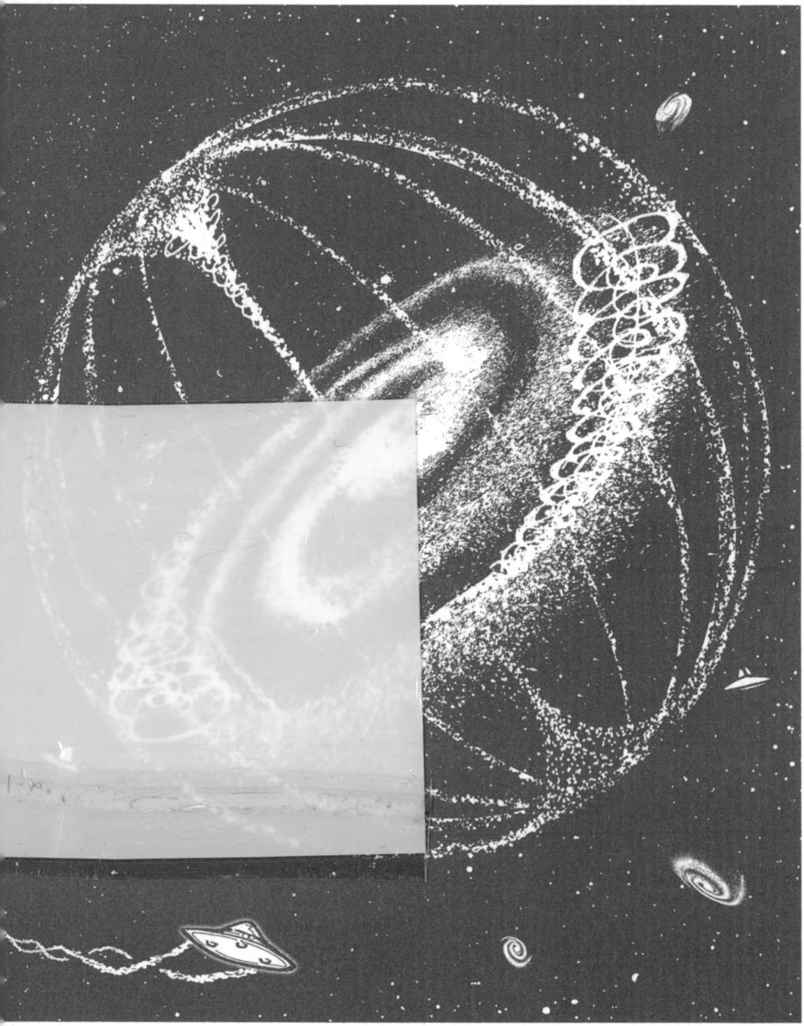

Time and Light Bubbles

all that we see

Our horizon, the limit of the visible, is a bubble expanding at the speed of light, about 186,000 miles per second. A light-year is the distance light covers in a year, a little under 5.9 trillion miles and a convenient measure for the vastness of space.

Light takes time to travel even at this incredible speed, so every glance can only peer into the past, and the more distant our gaze, the further back in time we see. Light from far quasars started off billions of years ago, back when the universe was (if you believe in the big bang) a sprightly youth. Much remains forever hidden, as illustrated by the boundaries of light cones (*opposite top*).

The theory of relativity links the three dimensions of space and the one of time into a single four-dimensional entity, space-time. We can imagine this as a stretchy sheet (*opposite center*). Anything placed on it distorts its shape; the more massive the object, the greater the distortion, affecting both space *and* time. It is this warping of space-time that causes the pull of gravitational fields and the closely related tug we feel when in accelerated motion.

Although we might not notice, everything is moving at the speed of light. We are hurtling through time at light-speed even if sitting reading a book. If we start to move through space, our velocity through time slows to make the combined space and time velocities still equal to the speed of light. At our usual slow pace the effects are minimal, however the faster we go the more noticeable this curious give and take becomes (*opposite below*).

All possible things that Star-Girl may see are contained within the past cone. Any event taking place outside is forever invisible as the light from it could never reach her eyes in the present moment.

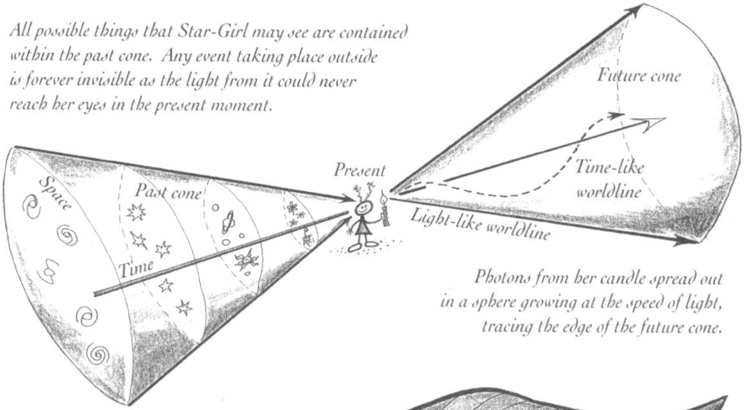
Future cone
Time-like
worldline
Present
Space
Past cone
Time
Light-like worldline

Photons from her candle spread out in a sphere growing at the speed of light, tracing the edge of the future cone.

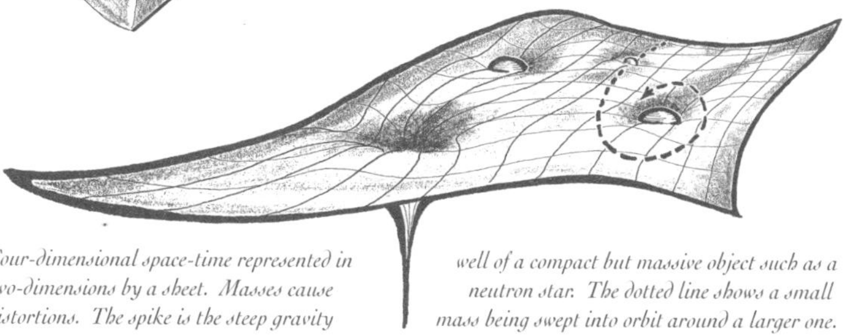

Four-dimensional space-time represented in two-dimensions by a sheet. Masses cause distortions. The spike is the steep gravity well of a compact but massive object such as a neutron star. The dotted line shows a small mass being swept into orbit around a larger one.

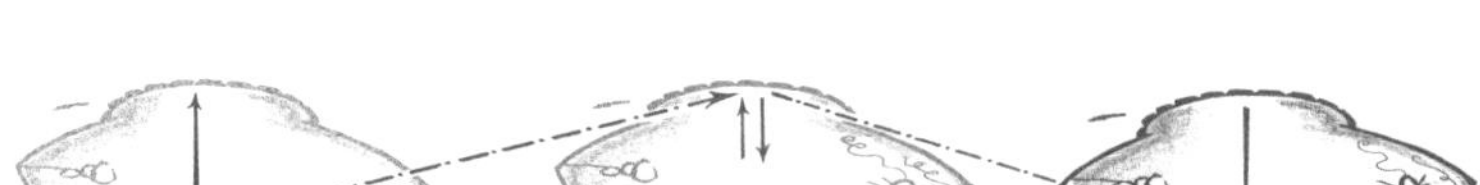

The Alien watches a photon bounce off the UFO roof. Star-Girl follows it as he flies by, seeing it apparently go farther.

Light cannot speed up, so it is time which has changed, passing more slowly in the spaceship than it is for stationary Star-Girl

Space-Time Games

there and back again

If we travel really fast, the relativistic effects become ever more severe as space and time juggle to conserve the speed of light. As an example, we shall send the Alien on a journey to a distant star, the UFO naturally flying at close to light-speed (1).

One of the first things Star-Girl notices is that the UFO appears shorter (2). This is a side effect of clocks running at different rates when in relative motion, since length from Star-Girl's perspective is measured by speed multiplied by time taken to go past a point. For the Alien, the craft's length remains the same.

As the UFO approaches light-speed, more and more energy is needed to accelerate, since rather than adding to the velocity some of the energy is instead converted into mass (3). This is a result of Einstein's $E=mc^2$ formula, which says that mass can be changed into an incredible amount of energy, and vice versa. In this instance the quicker an object travels the more total energy it has, thereby making it more massive and ever harder to accelerate; too close to the speed of light and mass shoots up to infinity, leaving an unfortunate Alien undergoing gravitational collapse and ending up as a black hole.

Luckily the Alien is wise, keeps his speed down, and manages the round trip to a nearby star in what seems a few months. Alas time has passed more slowly for the fast-moving traveler than for those left behind (4). On his return, all that remains of Star-Girl is her great great great great grandchildren (5), who grew up with an old family story of a friend who flew off to a distant star.

fig. 1
fig. 2
E=mc²
where E is energy
m is mass
c is the speed of light
fig. 3
fig. 4
RIP
RIP
RIP
RIP
fig. 5

Habitable Zones

life, the universe, and everything

This wonder we call life could well be a rare and precious thing. For earthly evolution we have needed a planet big enough to hold an atmosphere, close enough to the sun for water to be liquid, and yet far enough away to avoid an inferno.

The width of the comfortable circumstellar habitable zone has shrunk over time (*top left*) as the sun has become brighter. Alas, in a billion years or so the Earth itself will be too hot to sustain life as the sun continues its expansion into a red giant.

Our position in the galactic disk (*top right*) is a crucial balancing act. The planet and all its living bodies contain and require heavier elements synthesized in supernovae, yet the explosions of these dying suns sterilize enormous areas of space. Life struggles to develop near these fearsome events, making the still-active galactic center a no-go zone. Conversely, the outer edge of the disk has too few heavy elements, making life there also unlikely.

Should a species survive long enough to form an intelligent society, there are several stages to look forward to (*main opposite*). *Type 0* civilizations are divided by squabbles and inefficient use of resources. Rising to *Type I* status requires planetary consensus, global communication, and renewable energy sources.

Colonizing its solar system allows a *Type II* society to spread to neighboring planets and moons, so outliving its mother planet. Interstellar travel grants the next giant leap to a galactic *Type III* civilization, while eventually, with resources on a cosmic scale, *Type IV* beings can manipulate the very essence of space-time.

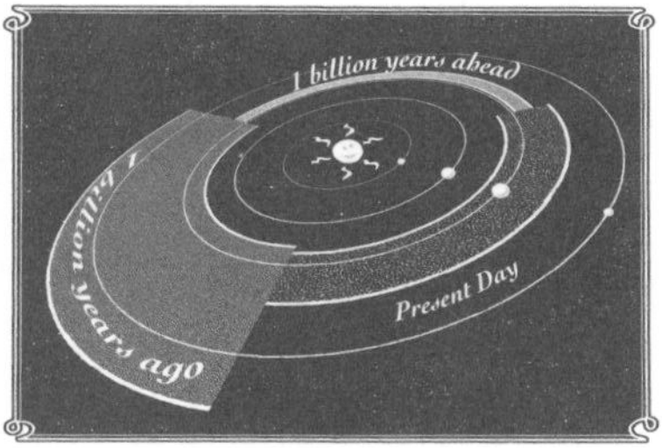
1 billion years ahead
1 billion years ago
Present Day

Circumstellar habitable zone

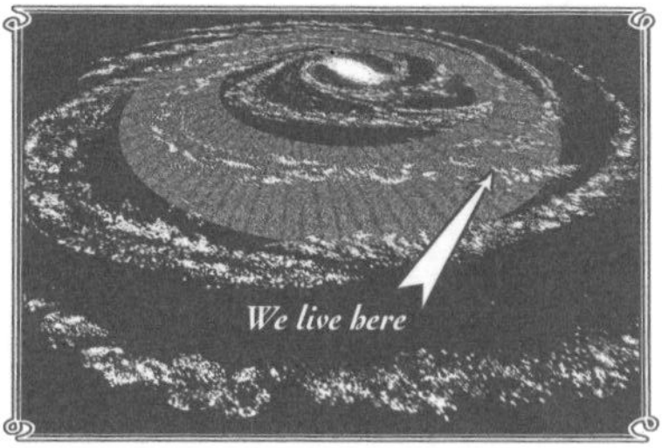
We live here

Galactic habitable zone

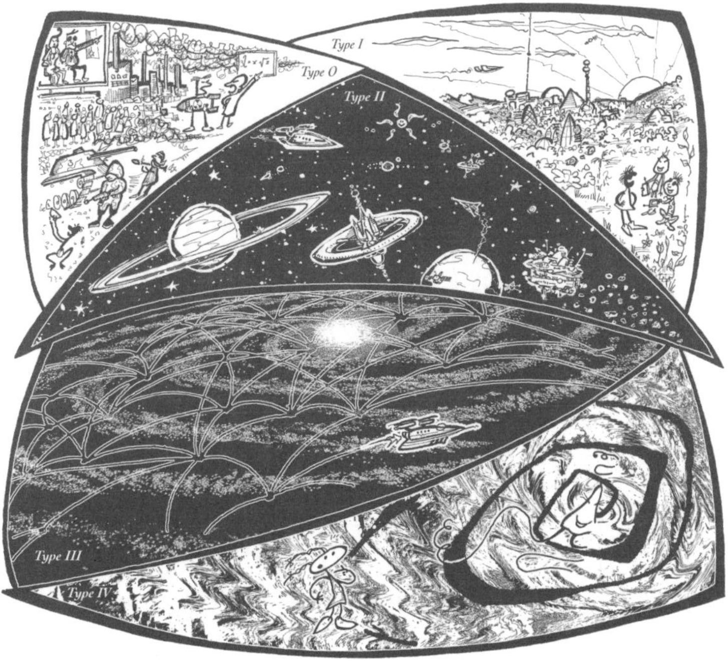
Type I
Type 0
Type II
Type III
Type IV

The evolution of civilization

Fine Tuning & Multiverses

through the wormhole

For us, the universe is just right. The anthropic principle takes this a step further by stating that if it wasn't, we wouldn't be here. One example of such cosmic benevolence is a convenient energy level of carbon which allows significant quantities to form in stars. Without it carbon-based life would be stuck for raw materials.

The behavior of matter and energy pivots on twenty or so finely-tuned constants, including the masses of fundamental particles, the strengths of the forces binding atomic nuclei, and the relative weakness of gravity compared with the other forces.

If these constants were varied even slightly then the universe would be a very different place. Atoms might be so unstable that matter itself could not exist, or stars would evolve unable to build the heavier elements needed for life. Perhaps the universe might not even get beyond an initial miniscule fireball.

Some theories allow for parallel universes, which can have very different rules. It could be that some or all possibilities, including ourselves, may exist as players in a larger multiverse.

Joining all this together, like holes in a Swiss cheese, could be the task of wormholes, highly speculative tunnels burrowing through the

fabric of space-time to link normally widely-separated regions. Creating a long-lived wormhole big enough to allow safe passage would probably require exotic matter and feats of engineering a trifle beyond our current feeble technologies.

Carbon-based life needs a specific multistage fusion reaction to occur in the stars.

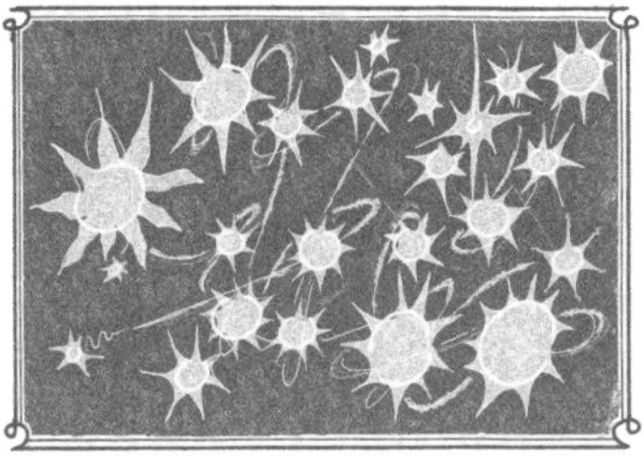

If gravity was stronger, galaxies would be densely packed with stars in chaotic orbits.

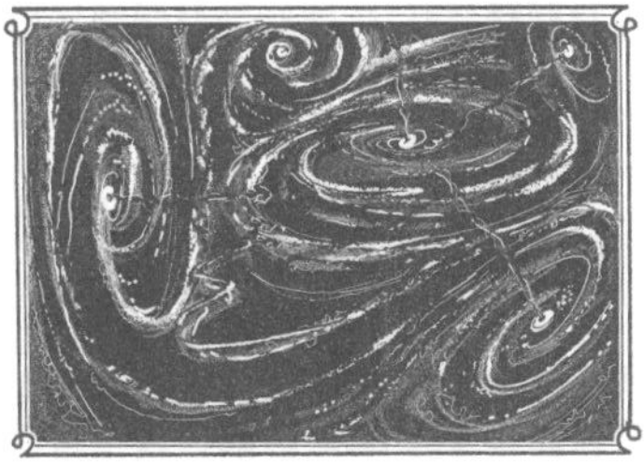

Huge black holes would result from a less uniform underlying universal structure.

Equal quantities of matter & antimatter would leave radiation and little else.

In an infinite multiverse all variations of universe would exist side by side, possibly interconnected by a network of wormholes.

Ultimate Reality
the kaleidoscopic playpen

With each passing second our view of the universe grows slightly wider, yet we still have little idea of how big it is, whether it has an edge, or what molds the structures we see. There is also the awesome possibility that ours may be one of an infinitude, with quite a few more dimensions than expected.

Are we alone, and if not, who and what might our neighbors be? Is a homely planet necessary, or can evolution act in more extreme niches? On the smallest scales, the distinction between matter and energy blurs, leaving seemingly solid reality a phantasm woven from mainly nothing. Some subatomic systems share a curious entanglement even if widely separated by space or time. Could this be the beginnings of consciousness? Maybe life is part of a greater series of interconnections, perhaps itself subtlely guided by an underlying panpsychic awareness.

Stranger still, we may be just scintillating illusions, extrapolated from deeper realities. Akin to photographic holograms that encode three-dimensional pictures onto flat film, our universe may arise from interference patterns on a hyperspatial boundary, interpreted by its enraptured inhabitants as a solid entity.

Humanity has only just peeped out from beneath its earthly blindfold and taken the first, stumbling steps off-world onto a marvellous odyssey. As a species we are facing many challenges, not least the ones we have created, yet our very atoms were born in the stars. One day perhaps we might return to visit.

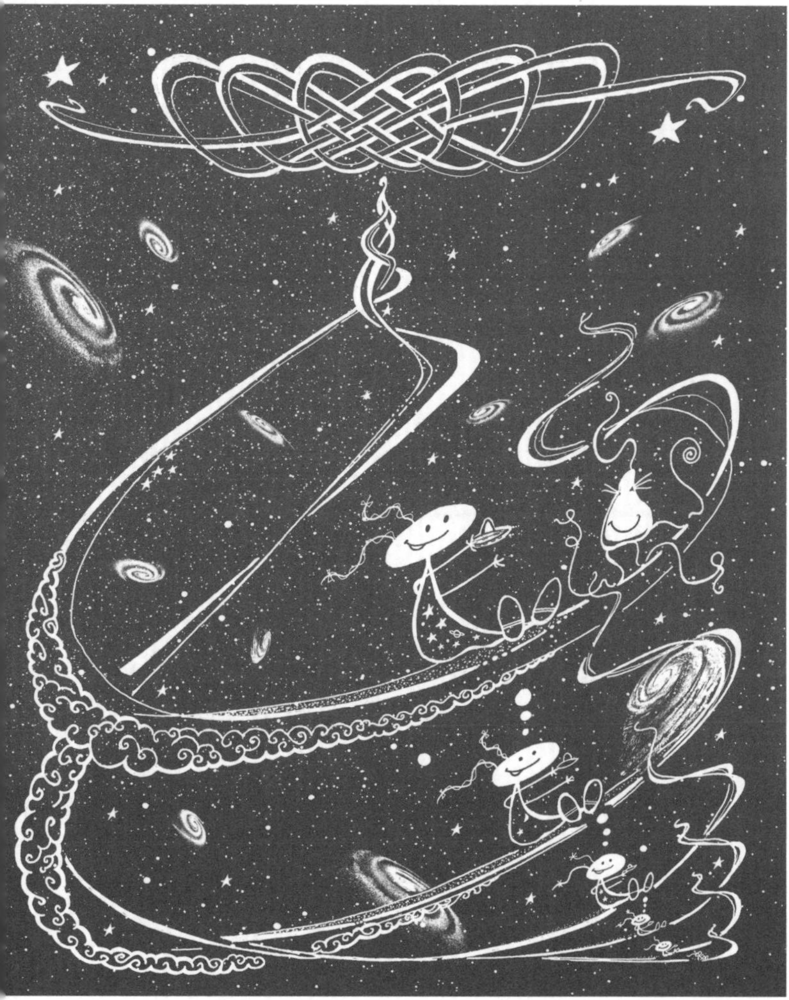

Northern Sky

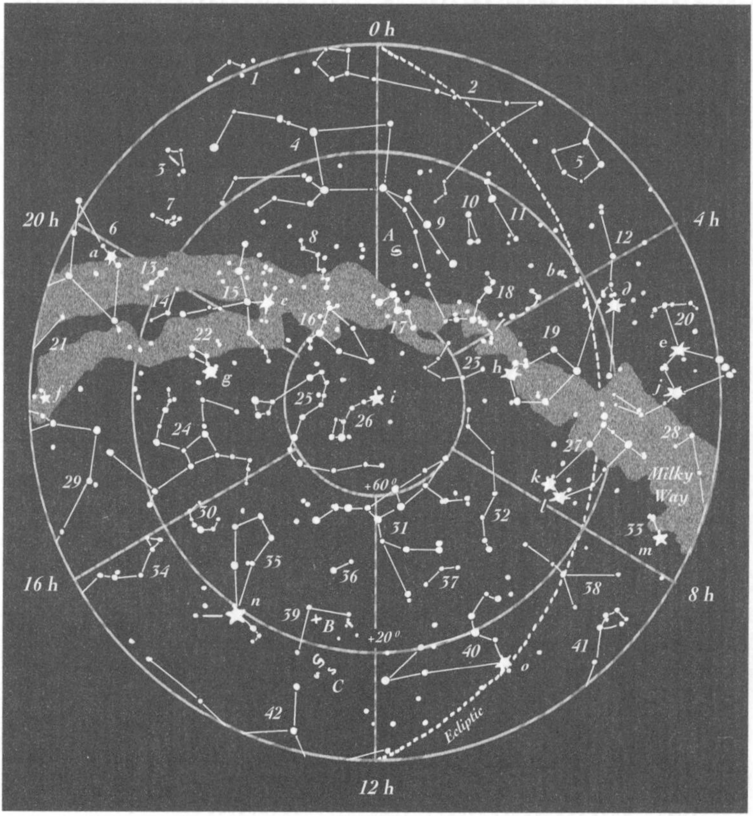

0 h
4 h
8 h
12 h
16 h
20 h
+60°
+20°
Milky Way
Ecliptic

Stars : *a*. Altair, *b*. Pleiades, *c*. Deneb, *d*. Aldebaran, *e*. Bellatrix, *f*. Barnard's Star, *g*. Vega, *h*. Capella, *i*. Polaris, *j*. Betelgeuse, *k*. Castor, *l*. Pollux, *m*. Procyon, *n*. Arcturus, *o*. Regulus. *Misc* : *A*. Andromeda galaxy, *B*. North galactic pole, *C*. Coma and Virgo galaxy clusters.

Constellations : 1. Aquarius, 2. Pisces, 3. Equuleus, 4. Pegasus, 5. Cetus, 6. Aquila, 7. Delphinus, 8. Lacerta 9. Andromeda, 10. Triangulum, 11. Aries, 12. Taurus, 13. Sagitta, 14. Vulpecula, 15. Cygnus, 16. Cephus, 17. Cassiopea, 18. Perseus, 19. Auriga, 20. Orion, 21. Serpens Cauda 22. Lyra, 23. Camelopardalis, 24. Hercules, 25. Draco, 26. Ursa Minor, 27. Gemini, 28.Monoceros, 29. Ophiuchus, 30.Corona Borealis, 31. Ursa Major, 32. Lynx, 33. Canis Minor, 34. Serpens Caput, 35. Bootes, 36. Canes Venatici, 37. Leo Minor, 38. Cancer, 39. Coma Berenices, 40. Leo, 41. Hydra, 42. Virgo

Southern Sky

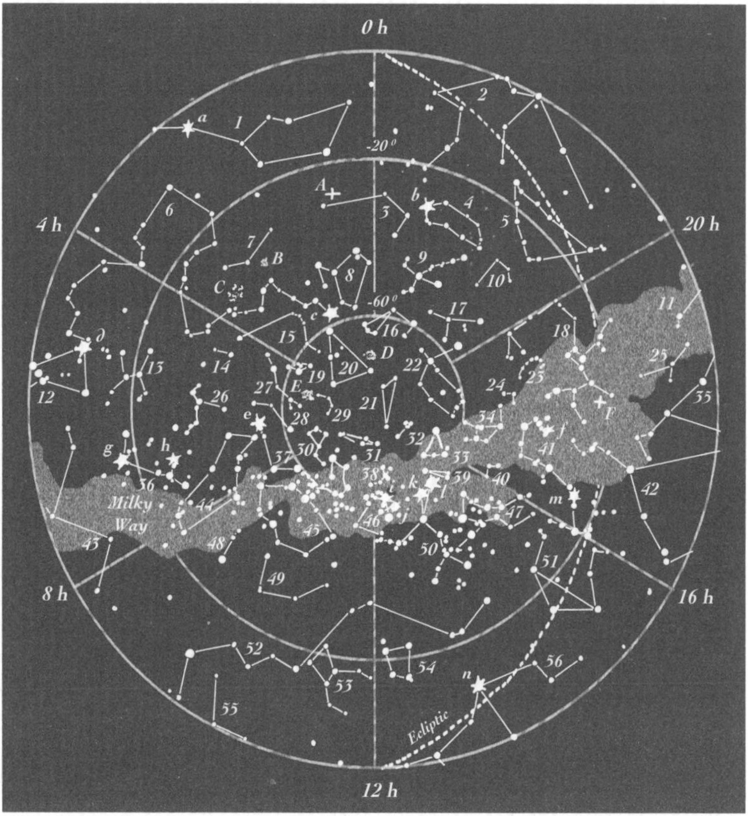
0 h
4 h
20 h
8 h
16 h
12 h
-20°
-60°
Milky Way
Ecliptic

Stars : *a*. Mira, *b*. Formalhaut, *c*. Achernar, *d*. Rigel, *e*. Canopus, *f*. Shaula, *g*. Sirius, *h*. Adhara, *i*. Acrus, *j*. Mimosa, *k*. Hadar, *l*. Rigel Kentaurus, *m*. Antares, *n*. Spica. *Misc* : *A*. South galactic pole, *B*. Fornax galaxy sys. *C*. Fornax galaxy clus. *D*. Small Magellanic cloud. *E*. Large Magellanic cloud. *F*. Galactic center

Constellations : 1. Cetus, 2. Aquarius, 3. Sculptor, 4. Piscis Austrinus, 5. Capricornus, 6. Eridanus, 7. Fornax, 8. Phoenix 9. Grus, 10. Microscopium, 11. Aquila, 12. Orion, 13. Lepus, 14. Caelum, 15. Horologium, 16. Tucana, 17. Indus, 18. Sagittarius , 19. Reticulum, 20. Hydrus, 21. Octans, 22. Pavo, 23. Corona Australis, 24. Telescopium, 25.Scutum, 26. Columba, 27. Dorado, 28. Pictor, 29. Mensa, 30. Volans, 31. Chameleon, 32. Apus, 33. Triangulum Australae, 34. Ara, 35. Serpens Cauda, 36. Canis Major, 37. Carina, 38. Musca, 39. Circinus, 40. Norma, 41. Scorpius, 42. Ophiuchus, 43. Monoceros, 44. Puppis, 45. Vela, 46. Crux, 47. Lupus, 48. Pyxis, 49. Antilia, 50. Centaurus, 51. Libra, 52. Hydra, 53. Crater, 54. Corvus, 55. Sextans, 56. Virgo

GALACTIC MAPS

superclusters, the local supercluster and the local group

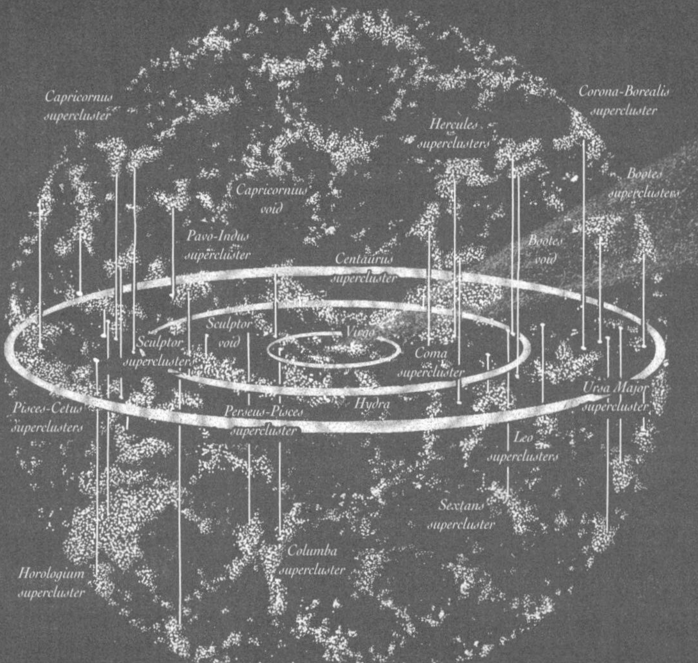

Capricornus supercluster
Corona-Borealis supercluster
Hercules superclusters
Capricornius void
Bootes superclusters
Pavo-Indus supercluster
Bootes void
Centaurus supercluster
Sculptor void
Sculptor supercluster
Virgo
Coma supercluster
Hydra
Ursa Major supercluster
Pisces-Cetus superclusters
Perseus-Pisces supercluster
Leo superclusters
Sextans supercluster
Columba supercluster
Horologium supercluster

Superclusters
rings are spaced 400 million light-years apart

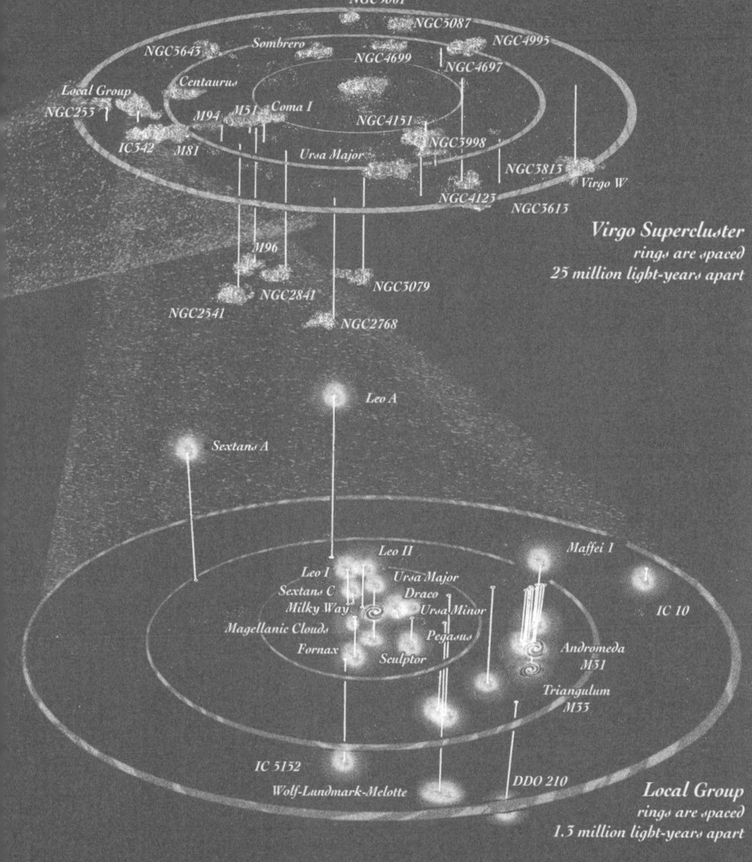
NGC5061
NGC5087
NGC4995
NGC5643
Sombrero
NGC4699
NGC4697
Centaurus
Local Group
NGC253
M94
M51
Coma I
NGC4151
IC342
M81
Ursa Major
NGC3998
NGC3813
Virgo W
NGC4123
NGC3613
M96
NGC3079
NGC2841
NGC2541
NGC2768
Virgo Supercluster
rings are spaced
25 million light-years apart
Leo A
Sextans A
Leo II
Maffei 1
Leo I
Ursa Major
Sextans C
Draco
IC 10
Milky Way
Ursa Minor
Magellanic Clouds
Pegasus
Fornax
Sculptor
Andromeda
M31
Triangulum
M33
IC 5152
DDO 210
Wolf-Lundmark-Melotte
Local Group
rings are spaced
1.3 million light-years apart

MISC DATA

Speed of Light	c	670,616,422 mph, 186,282.4 miles s^{-1}, 299 792 458 ms^{-1}
Light year	ly	0.306595 pc, 5,878,502,843,522 miles, 9,460,536,207,068 km
Parsec	pc	3.261 ly, 1.9173508×10^{13} miles, 3.085678×10^{13} km
Astronomical Unit	AU	1.5813×10^{-5} ly, 92,955,791 miles, 149,597,892 km
Earth Grav. Accel.	g_n	32.174 feet s^{-2}, 9.80665 ms^{-2}
Earth Escape Veloc.	v_{esc}	6.96 miles s^{-1}, 11.2 kms^{-1}

Earth Std. Atmos.		101,325 Pa
Atomic mass unit	u	1.66054×10^{-27} kg
Atm. mass eng. eqv.	$m_u c^2$	931.494 MeV
Avogadro's no.	N_A	6.02214×10^{23} mol^{-1}
Bohr radius	a_0	5.291772×10^{-11} m
Boltzmann const.	k	1.38065×10^{-23} $J K^{-1}$
Electron Mass	m_e	$9.1093826 \times 10^{-31}$ kg
		= 0.5110 MeV
Electron charge	e	1.602189×10^{-19} C
		= 4.8030×10^{-10} esu
Proton mass	m_p	$1.6726217 \times 10^{-27}$ kg
		= 1836.15 x electron mass
Neutron mass	m_n	1.674927×10^{-27} kg
Faraday constant	F	9.64853×10^{4} C mol^{-1}
Fine struct. const.	□	7.29735×10^{-3}
Gravitational const.	G	6.6742×10^{-11} m^{-3} kg^{-1} s^{-2}
Imped. of vacuum	□$_0$	376.7303 □
Loschmidt const.	n_0	6.6867775×10^{25} m^{-3}
Mag. flux quantum	□$_0$	$2.0678336 \times 10^{-15}$ Wb
Magnetic const.	□$_0$	4□ x 10^{-7} NA^{-2}
Permittivity const.	□$_0$	8.8542×10^{-12} F m^{-1}
Planck constant	h	$6.6260693 \times 10^{-34}$ Js
Planck length	l_P	1.616×10^{-35} m
Planck time	t_P	5.319×10^{-44} s

Planck mass	m_P	2.17645×10^{-8} kg
Planck temp	T_P	1.41679×10^{32} K
Steph-Boltz. const.	□	5.67040×10^{-8} W m^{-2} K^{-4}
Rcp. fine str. const.	1/□	137.036
Rydberg const.	R_H	1.097373×10^{7} m^{-1}
Wien disp. const.	b	2.8977685×10^{-3} m K
1st radiation const.	c_1	3.741771×10^{-16} W m^2
2nd radiation const.	c_2	1.438775×10^{-2} m K

Lyman series	*(ultraviolet)*	n □1	□□□*912 Å*
		n □2	□□□*1216 Å*
		n □3	□□□*1026 Å*
		n □4	□□□*973 Å*
Balmer series	*(visible)*	n □2	□□□*3650 Å*
		n □3	□□□*6560 Å*
		n □4	□□□*4860 Å*
		n □5	□□□*4340 Å*
		n □6	□□□*4100 Å*
Paschen series	*(infrared)*	n □3	□□□*8210 Å*
		n □4	□□□*18,761 Å*
		n □5	□□□*12,830 Å*
Brackett series	*(infrared)*	n □4	□□□*14,592 Å*
		n □5	□□□*40,532 Å*
		n □6	□□□*26,300 Å*

log_{10} metres -20 -15 -10 -5 0 5 10 15 20 25

Quarks Nucleus Molecule Sand Mouse Whale Asteroid Star Solar System Galaxy Galactic Wall

Proton Atom Cell Insect Human Forest Planet Super Giant Glob. Cluster Super Cluster